Guide to

LIVING MAMMALS

Previously published titles in this series
Guide to Invertebrate Animals (2nd Edition)
Guide to Living Reptiles
Guide to Living Birds

Forthcoming titles include
Guide to Fishes
Guide to Amphibians

Guide to
LIVING MAMMALS

J. E. WEBB
*Professor of Zoology, Westfield College,
University of London*

J. A. WALLWORK
*Reader in Zoology, Westfield College,
University of London*

J. H. ELGOOD
*Formerly Associate Professor of Zoology,
University of Ibadan, Nigeria*

Second Edition

First edition 1977
Second edition 1979

Published 1979 by
THE MACMILLAN PRESS LTD
London and Basingstoke
Associated companies in Delhi Dublin
Hong Kong Johannesburg Lagos Melbourne
New York Singapore and Tokyo

Printed and Bound in Great Britain by Bell and Bain Ltd.,
Glasgow

British Library Cataloguing in Publication Data

Webb, Joseph Ernest
    Guide to living mammals. − 2nd ed.
    1. Mammals − Classification
    I. Title   II. Wallwork, John Anthony
    III. Elgood, John Hamel
    599′.001′2   QL708

    ISBN 0 − 333 − 27257 − 9

Preface

This Guide to the Living Mammals is the second in a series intended to cover the animal kingdom and to provide, through an annotated classification of the groups, a basic understanding of the animals they contain. Like the first volume, Guide to Invertebrate Animals, most of the material and its mode of presentation has been used for a number of years in the training of students so that the mammal guide has evolved slowly as a result of this experience.

Although the concept developed in these books is the same, that is the learning of structure and relationships of animals through classification, they differ in a number of ways. In the mammals we are considering variety within a single class many of the members of which are familiar to most people. Among the invertebrates this was not the case. Here we had a wide range of organisms mostly known only to those with a special interest in animals and representing a number of fundamentally different types of structural organisation.

The mammals demonstrate perhaps better than any other group the effects of change in land form and climate on the distribution of animals. A major feature of this guide, therefore, is the inclusion of maps showing the areas occupied by all the important groups where distribution becomes significant. Such maps are, of course, only an approximation indicating the probable range of the wild forms. They do not give the present day distribution of domesticated and other mammals such as the horse, camel, rabbit and deer that have been spread by Man far beyond the areas where they are indigenous.

There are many books on the mammals, but three in particular supplement the brief treatment in this guide giving valuable additional information on classification, structure and distribution. The Principles of Classification and a Classification of Mammals by G.G. Simpson (Bull. Amer. Mus. nat. Hist. 85, 1945) is an excellent conspectus of the group. Recent Mammals of the World by S. Anderson and J. Knox Jones, Ronald Press, New York, 1967, is specially useful for the characters of the families and distribution, while the Traite de Zoologie by Grasse, volume 17 (in two parts), Masson, Paris, 1955, is an encyclopaedic treatment in French of the anatomy, behaviour and systematics of the mammalian orders.

The success of the mammal guide has encouraged us to extend the system of comparison by sets of matching characters to embrace all the mammalian families. This has been done in this second edition by including the families of marsupials, bats, lagomorphs, rodents, whales and sea cows which were omitted from the first edition. The coverage down to family level across the entire class is now comparable to that of the reptile guide and the bird guide shortly to appear and, we believe, much improves the value of the guide as a work of reference. Accordingly new lists of generic and common names have been prepared and cross-references within the text have been added.

We are grateful to Mrs. Margaret Clarke for the preparation of the typescript for photolithography and to Phil Brooks for the majority of the animal drawings. We are also indebted to Professor David Pye for much help with the bat families and to Mrs. Juliana Depledge for the bat drawings.

JEW
JAW
JHE
London, January 1979

Contents

C O N T E N T S

CONTENTS

C O N T E N T S

CONTENTS

1   Introduction

The mammals are of general interest because of all
animals they are the most like ourselves in behaviour and
structure and their lives most easily interpreted in human
terms. Man has always been closely associated with a variety
of mammals first as a hunter and consumer of wild animals,
using their skeletal parts and skins for implements, ornament,
clothing and shelter, and later as a stock raiser. The
remains of mammals recovered from excavations show this and
provide the anthropologist and archaeologist with much evidence
of the way of life in human settlements where there is no
written record.

But for the student and the teacher of zoology mammals
are particularly important. They represent a high level of
animal organisation and much of our knowledge of animal
physiology and behaviour relates to them. In ecology they
form a significant part of the biosphere and have a greater
effect than most animals in shaping the natural environment,
particularly in the warmer and drier parts of the world.

The Origin of the Mammals

The mammals arose from the therapsid or mammal-like
reptiles in the Triassic period and are recognised because
their jaw articulation is between the squamosal and
dentary bones and not the quadrate and articular bones as
in the reptiles. This somewhat arbitrary distinction
enables the fossil forms to be classified with the living
mammals which are more easily identified from hair,
mammary glands and other characteristic soft parts.
However it is probable that the squamosal-dentary jaw

1

articulation developed on several occasions so that, in this sense, the mammals are probably polyphyletic. Five groups of late Triassic mammals are generally recognised, largely from their teeth. The Multituberculata, a highly successful group of herbivores, had cheek teeth with longitudinal rows of cusps. They became extinct in the Eocene period. The Docodonta had molars with four cusps arranged in a square, the Triconodonta had cheek teeth with three cusps in a line and the Symmetrodonta and Pantotheria had triangular cheek teeth. These four groups became extinct in the Cretaceous.

Three mammalian groups exist today, first the Prototheria or monotremes represented by the duckbilled platypus and the spiny anteaters in Australasia, secondly the Metatheria or marsupials of Australasia and America, and, finally, the Eutheria or placental mammals with virtually world-wide distribution. It is a reasonable hypothesis that both the marsupial and the placental mammals came from common ancestry within the pantotheres in the first half of the Cretaceous. The monotremes, on the other hand, presumably have had a long, separate history, but how they originated is highly speculative since they are not known as fossils before the Pliocene. They are evidently the remnants of a very early group with many reptilian features both in structure and reproduction. For example the pectoral girdle is virtually reptilian and they lay leathery shelled eggs. Some aspects of skull structure suggest a relationship with the Docodonta, while a second view, based largely on the structure of the teeth in the duckbilled platypus, links the monotremes with the multituberculates. The tentative relationships between the living and extinct mammals are shown in the diagram on page 3.

The earliest Metatheria and Eutheria were small mammals resembling the opossums and the shrews respectively, which are themselves not very dissimilar and share quite a large range of primitive characters. Their fossil remains are fragmentary and often limited to the teeth and jaws so that again there is much uncertainty about their relationships. Such remains from the Cretaceous of Europe, Africa and Asia have been

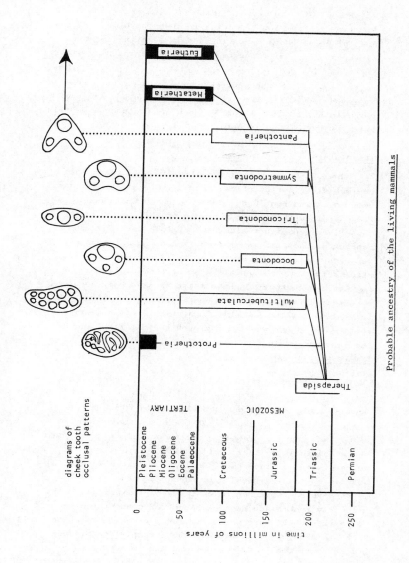

Probable ancestry of the living mammals

3

referred to the Eutheria. Other fossil material of
similar age from North America includes both metatherian
and eutherian characteristics and it seems that early
differentiation of these two lines was taking place in
North America in the lower and middle Cretaceous. In
South America, by the upper Cretaceous, the Metatheria had
achieved some diversity and there was also one group of
early Eutheria. On the other hand no Cretaceous mammals
have yet been found in Antarctica or Australia.

The Distribution of Marsupials

One of the problems of zoology is to explain the
present distribution of the marsupials. Australia has
been an island continent since the early Cretaceous, as
indeed was South America for much of this period, yet
these lands on either side of the globe have been the
centres of marsupial evolution. Land mammals are not
very good at crossing water and it is therefore tacitly
assumed that the marsupials must have walked into South
America and Australia, or at least crossed only a narrow
strait at a time when these continents were joined by a
third land mass, presumably Antarctica. (A summary of
continental drift showing the probable distribution of
the land masses in the Mesozoic is given in the Guide to
Living Reptiles.)

A second uncertainty is the effect of competition
with the placental mammals. The marsupials probably
became established before the placentals for a period of
time, negligible in geological terms, but sufficient to
allow the group to spread to all those regions they could
reach and in which they could survive. This is, of
course, surmise since the fossil record is quite inadequate
to give any real direction to the arguments. The
impressive adaptive radiation of marsupials in Australia
evidently took place in the absence of competition from
placentals, but there was a similar marsupial radiation
in South America, although all the marsupial carnivores
belonging to the Borhyaenoidea and many others are now
extinct. In this case, primitive placentals were either
present with the first marsupials or arrived a little

4

later and are today represented by the sloths, anteaters and armadillos. But there were other incursions into South America including the rodents and the primates in the Oligocene and much of the North American placental fauna in the Miocene when North America and South América became joined by the isthmus of Panama. Parallel with the marsupial carnivore radiation there was also an extensive placental herbivore radiation from the Palaeocene to the Miocene, including horse-like, camel-like, hippopotamus-like and elephant-like forms, all now extinct. Moreover, even now that North and South America are joined, marsupial opossums still remain in South America. In fact the so-called Virginian opossum, _Didelphis marsupialis_, has migrated into North America and has extended its range as far as southern Canada in recent time. The interaction between marsupials and placentals has clearly not resulted in the elimination of the marsupials in this case, though they were far more numerous before the junction with North America.

The Zoogeographical Regions

Whereas the marsupials radiated independently in South America and Australia, the placental mammals radiated mainly in North America, Europe, Asia and Africa including Madagascar, a vast area subdivided by seas, mountains and deserts and with climatic differences which changed considerably many times in the Tertiary.

The distribution of the mammals represents significant evidence on which the division of the globe into zoogeographical regions is based. Although the mammals as a whole, through their warm-bloodedness and diversity of habit, have colonised the great part of the earth's surface, particular groups have been isolated by geological and climatic events such as mountain building, continental drift, glaciation and hot desert formation. The distribution of species, families and sometimes orders tends to be determined by geographical corridors permitting their invasion of an area or by barriers that effectively halt their spread. According to their mode of life a geographical feature that may serve as a

corridor for one species may be a barrier to another and
vice versa. Thus a pattern of distribution emerges that
can only be understood from a knowledge of the past
history of the group, the geographical and climatic
changes that have occurred during that time, and the
influence of Man.

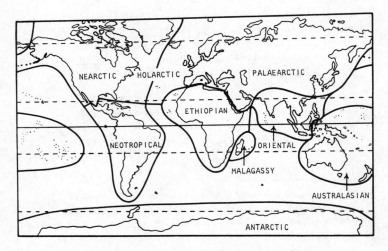

The zoogeographic regions of the World

(The Holarctic is a composite region comprising
the Palaearctic plus the Nearctic region)

The division of the land masses of the present into
regions each with a characteristic fauna was first broadly
recognised by P.L. Sclater in 1858 on the basis of the
distribution of birds. In the next decade these divisions,
with modifications, were found to apply also to the
reptiles and the mammals. In 1876 Alfred Russel Wallace
published Geographical Distribution of Animals and
considered the distribution of a much wider range of
vertebrate and invertebrate animals. He strongly

supported the idea of zoogeographical regions and these,
although further modified from time to time, have come
to be known as Wallace's Realms.   They are shown on the
accompanying map.   The faunal discontinuity between the
Oriental and Australasian regions is known as Wallace's
Line and passes between Bali and Lombok, between Borneo
and Celebes and runs eastward of the Philippines.

The Distribution of Placental Mammals

The zoogeographical regions owe the distinctness
of their fauna (and flora) to their degree of isolation,
that is to the barriers to dispersal of animals that have
arisen from time to time in the Tertiary.   Animal groups
that originated before this period have survived many
topographical changes in the lands and the seas and
have had the opportunity to colonise much of the Earth's
surface within the climatic belts to which their
physiology is adapted.   Thus many ancient groups are
either world-wide or occur within particular latitudinal
limits, or once had such a distribution, but have
disappeared in intervening areas leaving a discontinuous
pattern.   Among the mammals, the shrews (Family
Soricidae page 50) are ancient and virtually world-wide,
except for South America and Australasia.   The
emballonurid insectivorous bats, largely through powers
of flight, are pantropical in distribution (page 60).
Examples of discontinuous distribution are the tapirs
in tropical South America and in the tropical Far East
(page 200) and the camels and llamas in the Gobi desert
and in the Andes respectively (page 209).

The adaptive radiation of animals is best seen where
their dispersal is at a minimum and the products of
adaptation to different niches remain circumscribed by
barriers, as in islands isolated by wide stretches of
ocean.   Darwin's finches of the Galapagos Archipelago
are an example of this.   In Madagascar, an island
continent separate from Africa since the Cretaceous
Period, tenrecs (page 46), lemurs (page 82) and viverrid
carnivores (page 179) have radiated in isolation, and
form the greater part of the island's mammalian fauna
with a very high proportion of endemic species.   The

INTRODUCTION

hystricomorph rodents (page 130) in South America and
the marsupials (page 23) of Australia are similar cases.

Adaptive radiation is a normal occurrence in the
evolution of new groups or of animals colonising new areas
where there is a variety of unfilled niches.   The
resulting differently adapted forms then disperse as far
as barriers and corridors permit.   Invasion of new
territory does not necessarily result in the survival of
the species.   Climatic or other physical conditions may
not be suitable.   The invasion may not be in sufficient
force or with a frequency to permit a breeding population
to be established.   Competition with resident forms
better adapted to the existing conditions may prevent the
invaders gaining a 'footnold'.   But colonisation of a
new area takes place most readily where the endemic
populations have been eliminated or have suffered a
major setback from some physical effect such as a
change of climate or volcanic action.

Dramatic climatic changes in the Pleistocene have
been a major factor in determining the mammalian fauna
of the zoogeographical regions of the present.   In the
Miocene, mountain ranges arose extending from Spain to
the Malay peninsula and formed with the Mediterranean
and the Sahara, Arabian and Gobi deserts a barrier between
the Palaearctic, and the Ethiopian and Oriental regions.
The fauna of the Palaearctic, therefore, was isolated
except for a passage into the Oriental region in the east
through China and Malaya.   During the glacial periods
of the Pleistocene much of the Palaearctic was under ice
and the remainder suffered a much colder climate.   At
the same time the subtropics became wetter, the Sahara
and Arabian deserts were reduced and largely had a cover
of vegetation, and the increase in the polar ice lowered
the sea level.   Animals were driven south along the
eastern coastal belt of China and Malaya.   Land exposed
on the continental shelf of the Malay archipelago provided
a route to India and thence, with reduced deserts, to
Africa.   There was thus an influx of some Palaearctic
forms into the Oriental and Ethiopian regions, but not
all managed to pass.   There are, for example, almost no bears
(page 176), procyonids (page 175) or deer (page 213) in

8

the Ethiopian region. In the milder interglacial periods, rising sea level and expanding deserts again isolated the Ethiopian and Oriental regions and prevented a return of animals into the Palaearctic. The glacial-interglacial sequence thus caused progressive impoverishment of the Palaearctic mammalian fauna, an effect which was accentuated by the rise of Man. Similar events also took place in the New World and affected the distribution of mammals between the Nearctic and the Neotropical regions, but with far less reduction of the Nearctic fauna.

We are much concerned today with the preservation of animals threatened by over-exploitation or by the changes in environment caused by Man. Many of these endangered species are mammals, as might be expected, for the wild mammal tends to be in competition with Man himself for the prime requirements of food and a place to live and is either hunted or constrained to areas where survival is difficult.

There is ample need, therefore, for people to know more about mammals to provide the perspective essential not only for the zoologist but also for the ecologist, anthropologist, archaeologist, geographer and all those who feel concern for the environment. There is perhaps no better way of conveying information concisely than through the medium of classification.

Use of the Guide

In this book the classification of the mammals is taken not only as a means of recognition of the various groups to which they belong and hence of their probable relationships one with another, but as a basic framework for their structure, habit and distribution. To achieve this each group of mammals is compared with others at the same taxonomic level by sets of matching diagnostic characters. In this way the differences and similarities between orders within the class, or between families within an order are immediately apparent. The student has no difficulty in discovering for himself, why, for example, the various monkeys and lemurs belong to different families and suborders, but are all placed within the same order Primates

INTRODUCTION

and how this order may be recognised from other orders of mammals from which it differs in some respects but not in others. The various anatomical features used for this purpose are fully illustrated by numerous blackboard-style diagrams and the terms mentioned explained in the glossary. In addition the relationships of the groups are shown schematically, there are sketches of typical members of the groups and distribution maps of most of the orders and families. In this way an immense amount of basic information has been condensed into a small pocket book.

A great many of the mammals can be recognised and in fact classified by their teeth and their toes, in other words by what they eat and how they move. Within limits, therefore, and with the help of other external features, it is possible to gain a considerable insight into the structure and classification of the mammals in a tour of a zoo using this guide as an indication of what to look for. But this is not sufficient for all purposes. The student needs to know in some detail about internal structures, particularly skeletal structures, that are diagnostic of the various groups, while the archaeologist identifying mammal remains from an excavation inevitably must reach a conclusion as to their group entirely from skeletal material. Internal characters, mostly of the skull and the limb bones, therefore, are included in the diagnoses. In preparing the sets of matching characters it is obvious in a particular case that some will be of greater importance than others. These 'spot' characters that are of special importance in forming a diagnosis and sometimes unique to the group, such as the form of the limb in the artiodactyles or the structure of the teeth in the aardvark, are marked with a black spot.

In using the guide it will be found that illustrative diagrams of the various points are adjacent to the text or at least occur within the section dealing with a specific order of mammals. But some characters recur such as the type of gait, plantigrade, digitigrade and so forth, and are illustrated only once. Here reference to the term in the glossary gives not only a definition but the page on which the illustration occurs.

INTRODUCTION

The dental formula in general use throughout the guide refers to the complement of permanent teeth in the upper and lower jaws of one side of the head. The abbreviations I, C, Pm, M refer to incisor, canine, premolar and molar teeth respectively and the numbers, such as 2/3, refer to the number of teeth of a particular category in the upper/ lower jaw.

The student will find this guide helpful in the vertebrate and biogeography courses in a number of ways of which a few are listed below.

● It provides a conspectus of the mammals from which the range of diversity can be appreciated.

● Schematic diagrams show the basic classification in terms of the relationships thought to exist between the groups.

● The reasons for the classification are evident from the lists of matching characters. Here negative as well as positive characters are given and irrelevant features omitted.

● The diagrams and drawings are simplified giving only essential detail and for this reason are easy to remember.

● In the laboratory the guide serves as a reference book indicating the points for special note in demonstration specimens.

● The treatment lends itself to the construction of dichotomous keys.

● The guide forms a compact summary for revision purposes.

The preparation of lists of matching characters for the mammals does not appear to have been attempted on this scale before and has involved a considerable search in the literature with, in some cases, reference back to specimens. The classification used, while cognisant of modern views, has also to be one that works in these terms. Consequently some compromises have been made, not for the specialist, but for students for whom the guide is intended.

11

2 Egg-laying, marsupial and placental mammals

The mammals today are but a remnant of a once far more extensive and varied group. The living forms share common characters of their soft parts such as milk-producing mammary glands and hair, but these cannot be used for the many fossil species. For taxonomic purposes, therefore, the loss of the quadrate and articular bones to the middle ear, where they become the incus and the malleus, thus leaving the squamosal and the dentary to form the jaw articulation, becomes the essential character of the mammal.

Nevertheless the most significant mammalian feature is the evolution of the yolkless egg and the supply of nutrient to the embryo and foetus from the uterine wall culminating in the eutherian or placental condition. Evidence of the steps taken in this evolution comes from the monotremes, the duck-billed platypus and the spiny anteater, which still lay shelled eggs but feed their young on milk, and also from the marsupials.

In some marsupials the yolk sac of the embryo takes up a nutritive fluid (uterine milk) produced by the wall of the uterus and no direct connection between the embryo and the uterus is formed. But in others, for example in Dasyurus the marsupial cat, a yolk sac placenta develops, while in the bandicoot (Perameles) there is a temporary allantoic placenta. In every case retention of the embryo in the uterus is brief and the precociously developed young is soon transferred to the mammary nipple usually covered by a marsupium. The monotremes and marsupials are thus intermediate between the reptiles which lay shelled eggs but do not feed their young on milk and the Eutheria or placental mammals where an allantoic placenta is always present and the young are retained in the uterus up to an advanced stage of development.

Class Mammalia

Tetrapoda in which:-

● 1. The body is typically covered with hair. (14)

● 2. The head is on a flexible neck with typically seven cervical vertebrae and articulates through two occipital condyles. (16)

3. The brain is large and the cerebrum and the neopallium (associated with learned behaviour) tend to cover the mid-brain.

● There are four optic lobes (corpora quadrigemina). (42)

● 4. The dentary bone forms the lower jaw and articulates with squamosal. The quadrate and articular form auditory ossicles (incus and malleus) and the angular the tympanic bone. There is typically an external ear or pinna. (15)

5. Sound production is general and is achieved through vocal cords in the larynx, a modified region of the trachea. (15)

● 6. Teeth are typically present and are socketted (thecodont), differentiated (heterodont) and are replaced once (diphyodont) or are not replaced. (16)

● 7. The buccal cavity is enclosed laterally by cheeks and roofed by a false palate which separates it from the nasal cavity. (16)

8. The limbs tend to vertical orientation. (17)

● 9. Most of the long bones and vertebrae have epiphyses, bony caps separated from the main bone during growth by cartilage. In monotremes they are limited to the tail of Ornithorhynchus. (17)

●10. The heart is four-chambered giving rise to distinct systemic and pulmonary circulations. The right aortic arch is absent and the erythrocytes lack nuclei. (18)

●11. The thorax and abdomen are separated by a diaphragm. (18)

12. The egg is minute and develops in the uterus (except in monotremes which lay shelled eggs). The young remain with the

● female and are fed with milk from mammary glands. (14, 41)

13. A more or less high internal temperature is maintained.

N.B. see matching characters in guides to reptiles, birds etc.

CLASS	SUBCLASS
	PROTOTHERIA
MAMMALIA	METATHERIA
	EUTHERIA

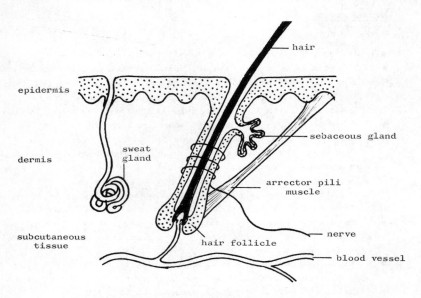

section through mammalian **skin**

primitive mammary area

mammary papilla

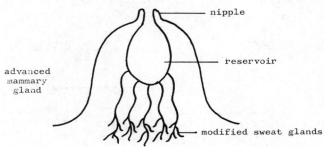

mammary glands

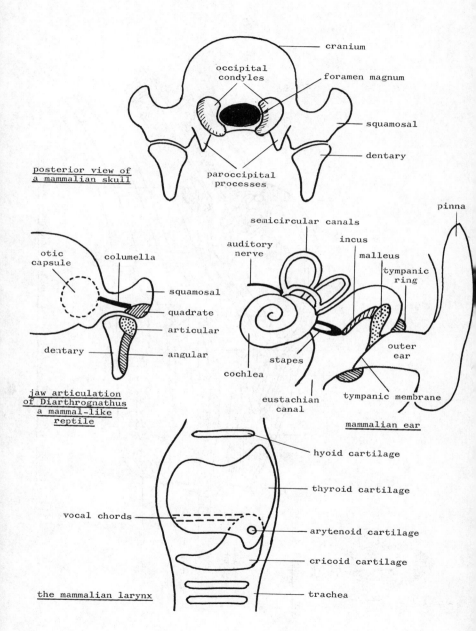

posterior view of
a mammalian skull

cranium

occipital
condyles

foramen magnum

squamosal

dentary

paroccipital
processes

pinna

semicircular canals

auditory
nerve

incus

malleus

tympanic
ring

otic
capsule

columella

squamosal

quadrate

articular

dentary

angular

stapes

cochlea

outer
ear

jaw articulation
of Diarthrognathus
a mammal-like
reptile

eustachian
canal

tympanic membrane

mammalian ear

hyoid cartilage

thyroid cartilage

vocal chords

arytenoid cartilage

cricoid cartilage

the mammalian larynx

trachea

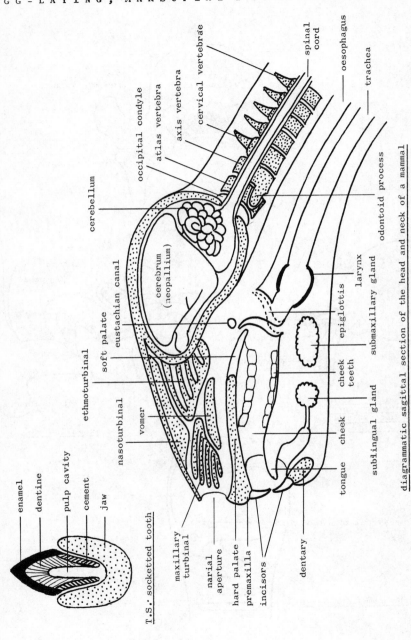

spinal cord

oesophagus

trachea

cervical vertebrae

axis vertebra

atlas vertebra

occipital condyle

cerebellum

odontoid process

cerebrum (neopallium)

larynx

submaxillary gland

eustachian canal

epiglottis

soft palate

cheek teeth

ethmoturbinal

cheek

nasoturbinal

sublingual gland

vomer

tongue

maxillary turbinal

narial aperture

hard palate

premaxilla

incisors

dentary

diagrammatic sagittal section of the head and neck of a mammal

enamel

dentine

pulp cavity

cement

jaw

T.S. socketed tooth

16

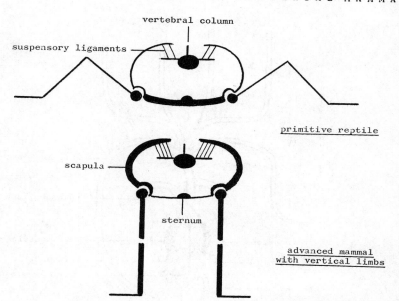

diagram showing the forelimbs and pectoral girdle of
a reptile and a mammal

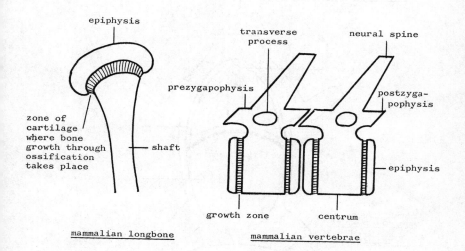

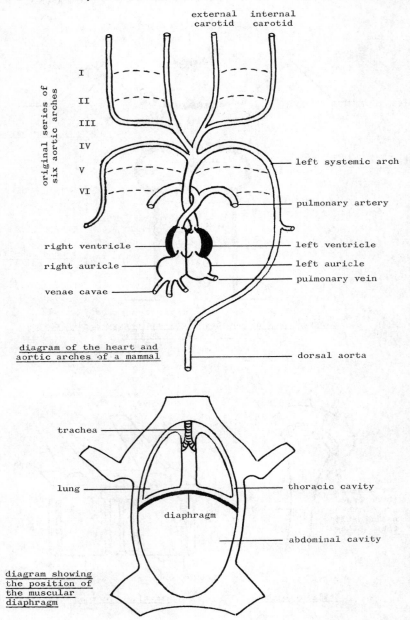

external carotid

internal carotid

original series of six aortic arches

I

II

III

IV

V

VI

left systemic arch

pulmonary artery

right ventricle

left ventricle

right auricle

left auricle

pulmonary vein

venae cavae

diagram of the heart and aortic arches of a mammal

dorsal aorta

trachea

lung

thoracic cavity

diaphragm

abdominal cavity

diagram showing the position of the muscular diaphragm

18

EGG-LAYING MAMMALS

Subclass Prototheria, Order Monotremata

Mammalia in which:-

- 1. The female lays large, yolky, shelled eggs.
- 2. The mammary glands do not have teats but open into a pair of longitudinal depressions. A temporary marsupial pouch supported by epipubic bones is present. (20)
- 3. A cloaca closed by a sphincter muscle is present into which open the urinogenital ducts and the rectum. The penis is attached to the floor of the cloaca, is bifid at the tip and serves only to carry sperm. The testes do not descend. (20)
  4. The ear is not protected by a bony covering or bulla and the tympanic bone forms an incomplete ring. There is no alisphenoid bone.
  5. The brain has no corpus callosum.
- 6. The ribs have a single head articulating with the centrum of the vertebra and are present in the cervical and thoracic regions and on some of the lumbar vertebrae.
  7. Teeth are present in <u>Ornithorhynchus</u> only, where they are small and calcified in the young animal, but are replaced by flattened, horny plates in the adult.
- 8. In the pectoral girdle the precoracoids are well developed and there is an interclavicle similar to that of the reptiles.
- 9. In the pelvis the ilium, ischium and pubis are not fused and there is a stout epipubic bone.
- 10. There is a poison spine on the hind tarsus of the male and a poison gland in the thigh. (20)

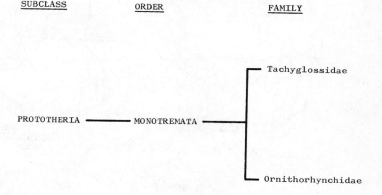

SUBCLASS	ORDER	FAMILY
PROTOTHERIA ——	MONOTREMATA ——	Tachyglossidae Ornithorhynchidae

19

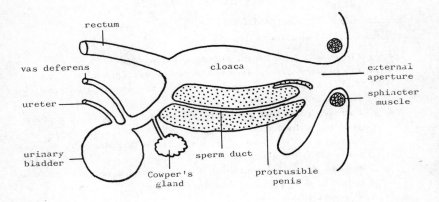

cloaca of a male monotreme

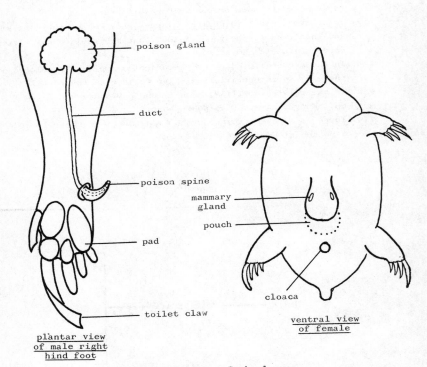

plantar view
of male right
hind foot

ventral view
of female

spiny anteater - Tachyglossus

Family Tachyglossidae – Echidna or spiny anteater

Monotremata in which:-

- 1. The body is covered with coarse hair and spines.
- 2. The snout is long and slender.
- 3. There is a long protractile tongue.
  4. The lower jaw is greatly reduced and there are no teeth.
- 5. The external ears are well developed but are partly concealed.
- 6. The limbs are modified for digging.(20)
  7. The brain is relatively large and the cerebral hemispheres convoluted.
  8. The food is termites, ants and other insects.
  9. They occur in Australia, Tasmania and New Guinea. There are 2 genera and 5 species.

Examples: Tachyglossus, Zaglossus.

spiny anteater
Tachyglossus

B

EGG-LAYING MAMMALS

<u>Family Ornithorhynchidae</u> - duckbilled platypus

Monotremata in which:-

● 1. The body is covered with soft hair.
● 2. The snout is broad and duck-billed.
  3. The tongue is flattened.
● 4. The lower jaw is stout and bears teeth. The dental formula is I 0/5, C 1/1, Pm 2/2, M 3/3.
  5. There are no external ears or pinnae.
● 6. The limbs are modified for swimming and digging.
  7. The brain is relatively small and the cerebral hemispheres smooth.
  8. The food is aquatic arthropods, molluscs and vegetation.
  9. They occur in all aquatic habitats in Eastern Australia and Tasmania. There is 1 genus and 1 species.

Example:- <u>Ornithorhynchus</u>.

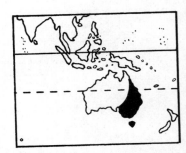

duckbilled platypus
<u>Ornithorhynchus</u>

MARSUPIALS

Subclass <u>Metatheria</u>, <u>Order Marsupialia</u>

Mammalia in which:-

1. The egg is yolky and covered with albumen and a membrane, but has no shell. It is retained within the uterus of the female. In some forms the yolk sac of the embryo absorbs nutritive fluid from the uterine wall, but in others (<u>Dasyurus</u>) there is a placental connection between the yolk sac and the uterus, and in <u>Perameles</u> a chorio-allantoic placenta is present.

● 2. The young are born at an early stage of development and in most species transfer to a marsupium enclosing distinct mammary teats.

3. There is a shallow cloaca with a common sphincter. Both the uterus and the vagina are double and the testes descend into scrotal sacs anterior to the penis. (25)

● 4. The ear is usually protected by a bony bulla formed from the alisphenoid. The tympanic bone is small, ring-shaped and not fused to the skull. (26)

5. The brain has no corpus callosum.

6. Movable ribs are restricted to the thorax. They are bicipital and articulate with the centrum and the transverse process of the vertebra.

7. There is one set of teeth which are not replaced except for the last premolar.

8. In the pectoral girdle the precoracoids are vestigial and the interclavicle absent. (27)

9. The ilium, ischium and pubis of the pelvic girdle are not fully fused and epipubic bones are frequently present. (27)

10. A poison spine is not present.

<u>N.B.</u> The marsupials have a number of primitive features in the skull which are also found in some Eutheria, particularly more primitive forms. Whereas no one of these alone is sufficient to characterize the group, together they provide a strong basis for diagnosis. The more important of these characters are given below.

● 11. The angle of the lower jaw has an inwardly directed shelf and is said to be inflected. (26)

● 12. The nasal bones are wide posteriorly. (26)

● 13. The premaxilla never touches the frontal bone. (26)

● 14. The jugal extends back to form part of the glenoid fossa. (26)

● 15. The posterior region of the palate usually has large vacuities and a thickened margin. (26)
● 16. The periotic has a backwardly directed mastoid process which appears on the surface of the skull between the squamosal and the exoccipital. (26)

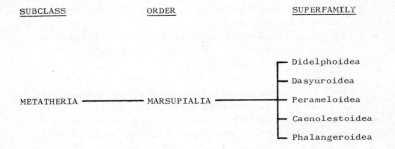

SUBCLASS	ORDER	SUPERFAMILY
METATHERIA	MARSUPIALIA	Didelphoidea Dasyuroidea Perameloidea Caenolestoidea Phalangeroidea

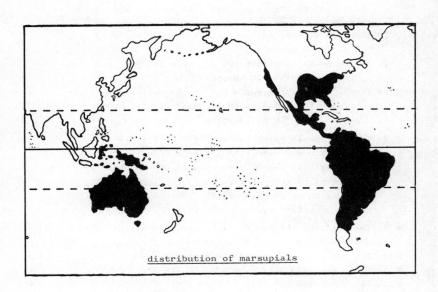

distribution of marsupials

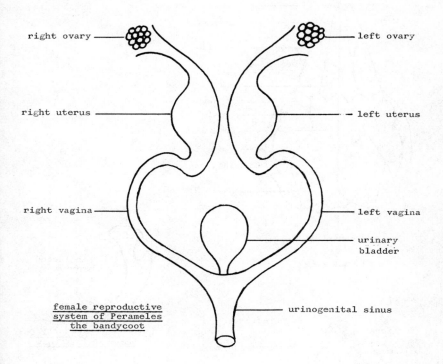

right ovary — left ovary

right uterus — left uterus

right vagina — left vagina

urinary bladder

<u>female reproductive system of Perameles the bandycoot</u>

urinogenital sinus

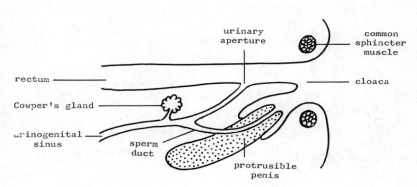

urinary aperture

common sphincter muscle

rectum — cloaca

Cowper's gland

urinogenital sinus

sperm duct

protrusible penis

<u>male reproductive system of Perameles the bandycoot</u>

25

MARSUPIALS

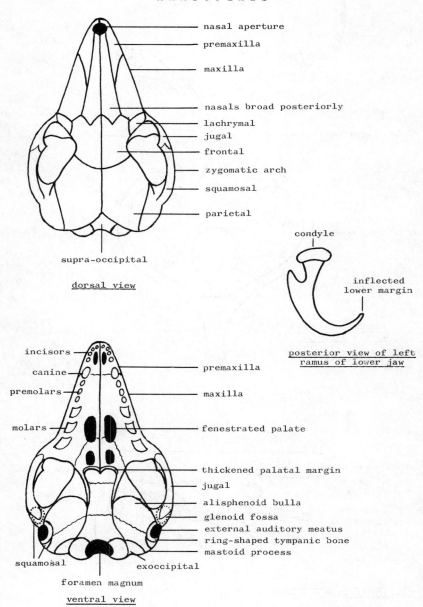

nasal aperture
premaxilla
maxilla

nasals broad posteriorly
lachrymal
jugal
frontal
zygomatic arch
squamosal
parietal

supra-occipital

<u>dorsal view</u>

condyle

inflected
lower margin

<u>posterior view of left
ramus of lower jaw</u>

incisors
canine
premolars

molars

premaxilla

maxilla

fenestrated palate

thickened palatal margin
jugal
alisphenoid bulla
glenoid fossa
external auditory meatus
ring-shaped tympanic bone
mastoid process

squamosal

exoccipital

foramen magnum

<u>ventral view</u>

<u>a polyprotodont marsupial skull</u>

26

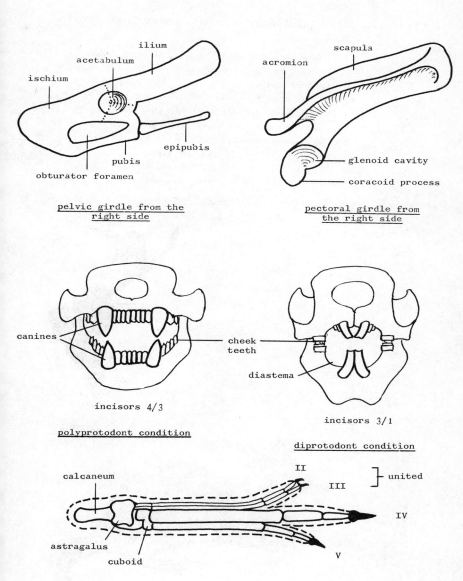

pelvic girdle from the
right side

pectoral girdle from
the right side

incisors 4/3

polyprotodont condition

incisors 3/1

diprotodont condition

syndactylous right foot

M A R S U P I A L S

<u>Superfamily Didelphoidea</u>, <u>Family Didelphidae</u> - Opossums

Marsupialia in which:-

- 1. There are more than three pairs of incisors of equal size in the upper and the lower jaws (polyprotodont). (27)

- 2. In the hind foot the 2nd and 3rd digits are not united (didactylous) and the 1st digit is opposable and clawless.

  3. The marsupium may be well developed, formed from two folds of skin, or absent.

  4. They are found in North and South America. There are 12 genera and 66 species.

  Examples:- <u>Didelphis</u>, <u>Chironectes</u>.

common opossum
<u>Didelphis</u>

distribution of
the Didelphidae

water opossum - <u>Chironectes</u>

MARSUPIALS

Superfamily Dasyuroidea

Marsupialia in which:-

● 1. There are three pairs or more of incisors of
equal size in the upper and the lower jaws
(polyprotodont). (27)

● 2. In the hind foot the 2nd and 3rd digits are not
united (didactylous) and the first, when present,
is clawless and not opposable, but usually
vestigial or absent.

3. The marsupium, when present, opens backward,
but is often only conspicuous in the breeding
season.

SUPERFAMILY	FAMILY
	┌─ Dasyuridae
Dasyuroidea ─────────────────┤
	└─ Notoryctidae

Family Dasyuridae - Marsupial carnivores, rats, mice, and anteater

Dasyuroidea in which:-

1. The pinna of the ear is present.

2. The eyes are well developed.

3. The forelimb is not modified for digging.

● 4. The incisors are small and pointed, the
canines large with a cutting edge and the
premolars and molars each have more than
one sharp cusp.

5. They are widespread in Australia, Tasmania,
New Guinea and the Aru and Normanby Islands.
There are 50 species in 20 genera.

Examples:- <u>Dasyurus</u>, <u>Dasyurops</u>, <u>Sarcophilus</u>,
<u>Thylacinus</u>, <u>Sminthopsis</u>, <u>Myrmecobius</u>.

(see pages 30 & 31)

Tasmanian devil
Sarcophilus

distribution of
the Dasyuridae

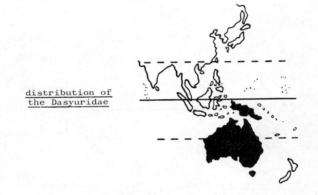

marsupial cat - Dasyurops

marsupial mouse - <u>Sminthopsis</u>

banded anteater - <u>Myrmecobius</u>

MARSUPIALS

Family Notoryctidae - Marsupial mole

Dasyuroidea in which:-

- 1. The pinna of the ear is absent.
- 2. The eyes are atrophied.
- 3. In the forelimb digits 3 and 4 have large triangular claws for digging.
- 4. The incisors, canines and premolars are simple and often blunt. The premolars and molars have one or two cusps. The teeth are well separated.
   5. They are found in South and north western Australia. There are two species in a single genus.

   Example:- Notoryctes

marsupial mole
Notoryctes

Superfamily <u>Perameloidea</u>, <u>Family Peramelidae</u> - Bandicoots

Marsupialia in which:-

● 1. There are three pairs or more of incisors of
equal size in the upper and the lower jaws
(polyprotodont). (27)

● 2. In the hind foot the 2nd and 3rd digits are
united in a common sheath (syndactylous) and
the first is reduced or absent. (27)

  3. The marsupium opens backward.

  4. They are found in Australia, Tasmania and
New Guinea.   There are 8 genera and 22 species.

   Examples:- <u>Perameles</u>, <u>Chaeropus</u>.

long-nosed bandicoot - <u>Perameles</u>

pig-footed bandicoot - <u>Chaeropus</u>

<u>distribution of
the Peramelidae</u>

Superfamily Caenolestoidea, Family Caenolestidae -

Rat opossums

Marsupialia in which:-

- 1. The first pair of lower incisors is large (diprotodont).(27)
- 2. In the hind foot the 2nd and 3rd digits are not united (didactylous) and the first has a claw, but is not opposable.
  3. There is no marsupium.
  4. They are found in the Andes of Venezuela, Colombia, Ecuador, Southern Peru and Chiloe Island.   There are 3 genera and 7 species.

   Examples:- Caenolestes, Lestoros, Rhyncholestes.

distribution of Caenolestoidea

rat opossum - Caenolestes

M A R S U P I A L S

Superfamily Phalangeroidea

Marsupials in which:-

● 1. The first pair of lower incisors is large
(diprotodont).(27)

● 2. In the hind foot the 2nd and 3rd digits are
united in a common sheath (syndactylous) and
the first is typically well developed and
opposable, but has no claw.(27)

3. The marsupium is well developed and opens
forward (except in Phascolarctos and Vombatus
where it opens backward).

SUPERFAMILY   FAMILY

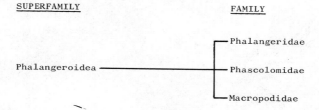

Phalangeroidea ─────────────── Phalangeridae

Phascolomidae

Macropodidae

Family Phalangeridae - Koala, Possums and Cuscuses

Phalangeroidea in which:-

● 1. There are two or three pairs of upper incisors.
● 2. The hallux is well developed and opposable.
● 3. The tail is usually long and often prehensile,
but vestigial in Phascolarctos.
4. They are widespread in Australasia.
There are 46 species in 16 genera.

Examples:- Phalanger, Trichosurus, Petaurus,
Phascolarctos, Tarsipes.

(see page 36)

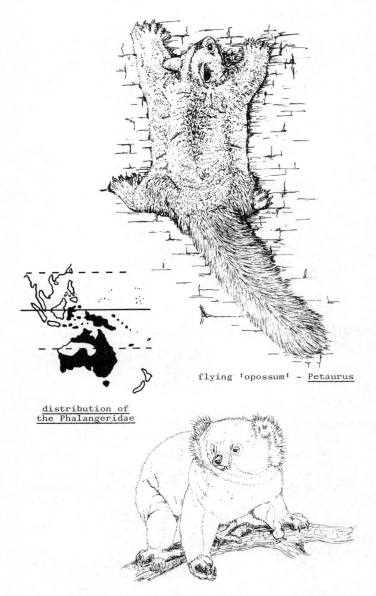

flying 'opossum' - <u>Petaurus</u>

distribution of
the Phalangeridae

koala bear - <u>Phascolarctos</u>

M A R S U P I A L S

Family Phascolomidae - Wombats

Phalangeroidea in which:-

● 1. There is a single pair of upper incisors.
● 2. The hallux is vestigial.
● 3. The tail is vestigial.
4. They are found in southeast Australia and Tasmania. There are two species in separate genera.

Examples:- Vombatus, Lasiorhinus.

distribution of the Phascolomidae

wombat - Vombatus

Family Macropodidae - Kangaroos and Wallabies

Phalangeroidea in which:-

● 1. There are three pairs of upper incisors.
● 2. The hallux is absent (except in Hypsiprymnodon).
● 3. The tail is long, but not prehensile.
4. They are widespread in Australasia (introduced into New Zealand and many small islands). There are 47 species in 19 genera.

Examples:- Macropus, Dendrolagus, Wallabia, Bettongia, Hypsiprymnodon.

(see page 38)

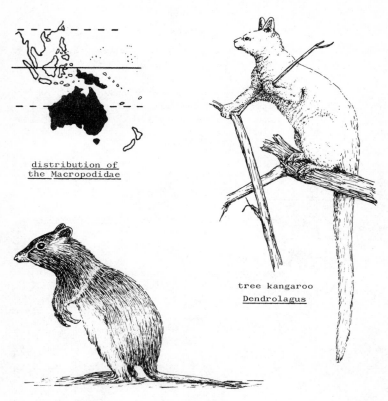

distribution of
the Macropodidae

tree kangaroo
Dendrolagus

muskrat kangaroo - Hypsiprymnodon

kangaroo - Macropus

Subclass Eutheria

Mammalia in which:-

1. The egg is very small with little or no yolk and a vitelline membrane. It is retained within the uterus. An allantoic placenta is present. (41)

● 2. The young are born at an advanced stage of development. There is no marsupium, the mammary teats projecting freely, in most forms, from the ventral surface of the body and often restricted to either the pectoral or the inguinal region. (14)

3. There is no cloaca and rarely a common sphincter except in a few primitive forms. The female ducts tend to become fused into a single structure so that there is always a median vagina and there may be a median uterus. In most forms the testes descend into scrotal sacs posterior to the penis. (14)

● 4. The middle ear is protected by a bony bulla except in the most primitive forms. Various bones may contribute to the bulla, but not the alisphenoid.

● 5. The cerebral hemispheres of the brain are connected by the corpus callosum. (42)

6. Movable ribs are restricted to the thorax. They are bicipital and articulate with the centrum and the transverse process of the vertebra.

7. There is typically a milk dentition of incisors, canines and premolars which are replaced by the permanent dentition which also includes molars. The basic formula is I 3/3, C 1/1, Pm 4/4, M 3/3.

8. In the pectoral girdle the coracoid is reduced to the coracoid process of the scapula and the interclavicle is absent. There is often a reduction or absence of the clavicle.

9. The ilium, ischium and pubis of the pelvic girdle become fused to form a single bone, the os innominatum, and there is no epipubis.

10. A poison spine is not present.

PLACENTAL MAMMALS

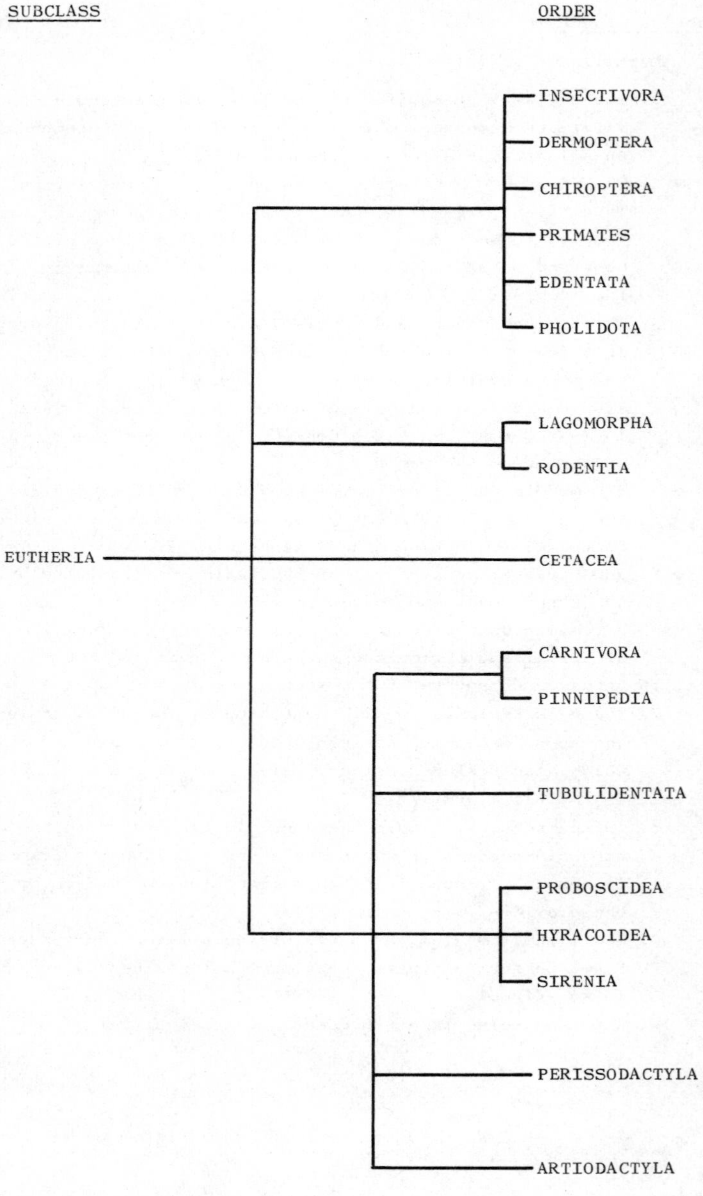

SUBCLASS

ORDER

INSECTIVORA

DERMOPTERA

CHIROPTERA

PRIMATES

EDENTATA

PHOLIDOTA

LAGOMORPHA

RODENTIA

EUTHERIA ——————————— CETACEA

CARNIVORA

PINNIPEDIA

TUBULIDENTATA

PROBOSCIDEA

HYRACOIDEA

SIRENIA

PERISSODACTYLA

ARTIODACTYLA

40

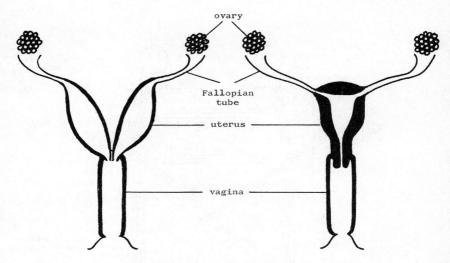

diagram to show the coalescence of the oviducts

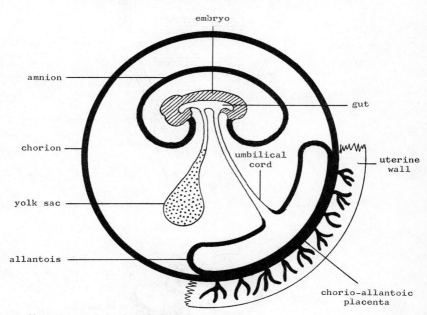

diagram to show the placental
connection between embryo and mother

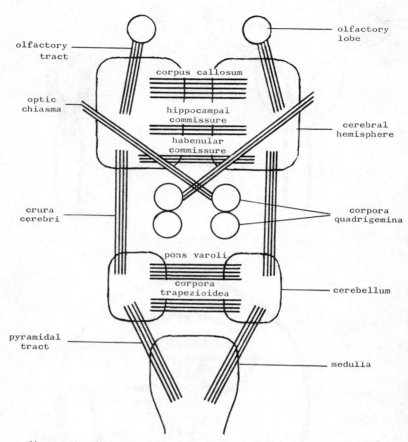

diagram to show the main connections in the eutherian brain

diagrammatic section through the cerebrum of a primitive eutherian

3 Insectivora, flying lemurs and bats

Although John Ray in the 17th Century with great foresight used toes and teeth as the basis for mammalian classification, he failed to recognise the Insectivora as a natural group. This was evidently due to the fact that the form of their toes and teeth, with some exceptions such as the limbs of moles, are generalised and do not show the special adaptive characters that have become diagnostic of the higher groups.

In fact the most primitive of the placental mammals, the Insectivora, and the primitive marsupials such as the opossums are not very different. Both are relatively small and live mostly retiring lives in well-wooded warm or temperate surroundings, feeding on small animals and eggs, which was presumably also the life style of the 'ancestral mammal'. Both also have retained a different but overlapping spectrum of primitive characters selected from those which it may be presumed were generally possessed by their ancestors. In consequence the insectivores have some primitive characters that the marsupials do not have and vice versa.

The two orders of flying mammals, the Dermoptera or flying lemurs and the Chiroptera or bats, are both ancient in origin and close to the insectivores. The flying lemurs are a small group of gliding mammals with limited Asiatic distribution. They probably represent an early stage in the evolution of powered flight through which the bats at some time passed. The bats on the other hand are a very large group with many different species. They show extreme adaptation to flight and are one of the most specialised groups of mammals. They are divided into two suborders, the more primitive flying foxes limited to the Old World tropics and subtropics and the more advanced insect-eating bats which are world-wide in distribution.

Order Insectivora

Eutheria in which:-

1. They are very small to small mammals terrestrial, burrowing, arboreal or amphibious, with typically a plantigrade gait, feeding mainly on invertebrates, eggs or small vertebrates. (45)
2. The limbs are of a generalised pentadactyl form with the digits ending in claws and in some cases modified for digging or swimming.
● 3. The teeth have sharp cusps associated with a carnivorous diet.
● 4. The olfactory region of the skull is longer than the cranial region.
5. There is no postorbital bar.
6. The lachrymal does not extend onto the face.
7. There is usually a distinct paroccipital process.

ORDER   FAMILY

```
                                              ┌─ Tenrecidae
                                    ┌─────────┤
                                    │         └─ Solenodontidae
                                    │
                                    ├──────────── Chrysochloridae
                                    │
INSECTIVORA ────────────────────────┼──────────── Erinaceidae
                                    │
                                    │         ┌─ Soricidae
                                    ├─────────┤
                                    │         └─ Talpidae
                                    │
                                    └──────────── Macroscelididae
```

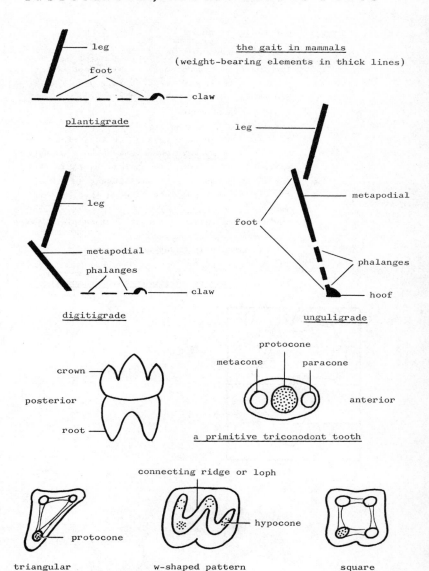

the gait in mammals
(weight-bearing elements in thick lines)

leg
foot
claw

plantigrade

leg
metapodial
phalanges
claw

digitigrade

leg
metapodial
foot
phalanges
hoof

unguligrade

crown
posterior
root

protocone
metacone
paracone
anterior

a primitive triconodont tooth

connecting ridge or loph
protocone
hypocone

triangular   w-shaped pattern   square

patterns on the occlusal surface of insectivore molars
(teeth with the cones joined by ridges or lophs are lophodont)

Family Tenrecidae - Tenrecs

Insectivora in which:-

1. They have undergone considerable adaptive radiation to give rise to forms resembling hedgehogs, shrews, moles and otters. A cloaca is present.
2. The limbs are variously modified.
● 3. The upper molars are triangular in surface view except in _Potamogale_ where they have a W-shaped pattern. The dental formula is variable, but
● the first premolar is always absent.(45)
● 4. The zygoma is incomplete through loss of the jugal.
5. The auditory bulla is absent or incomplete and the tympanic ring-shaped.
6. They are found in Madagascar and West Africa. There are 11 genera and 23 species.

Examples:- _Tenrec_, _Oryzorictes_, _Potamogale_

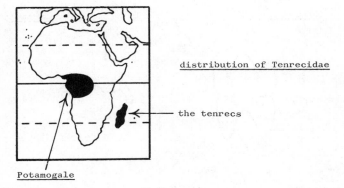

distribution of Tenrecidae

the tenrecs

Potamogale

common tenrec - _Tenrec_

46

Family Solenodontidae - Solenodons

Insectivora in which:-

1. They have small eyes and ears and a long scaly
   tail.   The hair is coarse and bristly.
2. The limbs have five digits with long, strong
   claws.
● 3. The upper molars are triangular in surface view.(45)
   The dental formula is I 3/3, C 1/1, Pm 3/3, M 3/3.
● 4. The zygoma is incomplete through loss of the
   jugal.
5. The tympanic is ring-shaped and there is no
   auditory bulla.
6. They are found in Cuba and Haiti.   There
   are 2 genera and 2 species.

Example:- Solenodon , Atopogale.

distribution of Solenodontidae

Cuban solenodon - Atopogale

Family Chrysochloridae - Golden Moles

Insectivora in which:-

- 1. The eyes are covered with skin and there is no external ear. The tail is rudimentary. The fur has a metallic sheen.
- 2. The fore-limbs and hands are highly modified for digging. The fifth digit is absent and the second and third bear large pick-like claws.
- 3. The upper molars are triangular in surface view.(45) The dental formula is I 3/3, C 1/1, Pm 3/3, M 3/3.
- 4. A secondary zygoma is formed from the maxilla and the squamosal and sometimes the parietal. The jugal is absent.
  5. The tympanic forms a prominent auditory bulla.
  6. They are found in tropical and South Africa. There are 5 genera and 11 species.

Examples:- Chrysochloris.

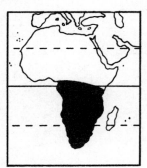

distribution of Chrysochloridae

the Cape golden mole
Chrysochloris

Family Erinaceidae - Hedgehogs

Insectivora in which:-

- 1. The eyes and ears are well developed. There is a tail of variable length and the body is covered with spines (Erinaceinae) or hair (Echinosoricinae).

  2. The limbs have five digits except in Atelerix which has four in the foot.

- 3. The upper molars are square in surface view.(45) The dental formula is I 2-3/3, C 1/1, Pm 3-4/2-4, M 3/3.

- 4. There is a strong zygoma formed from the maxilla, jugal and squamosal.

  5. The auditory bulla is incomplete and the tympanic bone ring-shaped and not fused to the cranium.

  6. They occur in Europe, Africa and Asia. There are 10 genera and 14 species.

  Examples:- Erinaceus, Atelerix, Echinosorex, Podogymnura.

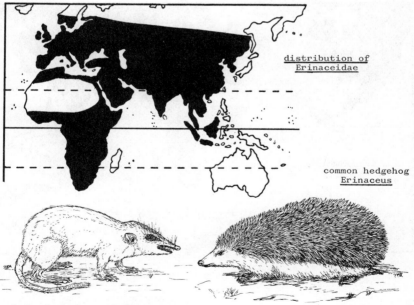

distribution of
Erinaceidae

common hedgehog
Erinaceus

hairy hedgehog - Podogymnura

Family Soricidae - Shrews

Insectivora in which:-

1. The eyes are small and the external ears usually visible. The tail may be long or short. The fur is short and thick. They include the smallest mammals.

2. The limbs have five digits.

● 3. The upper molars have a W-shaped pattern. (45) The first upper incisor is long and sickle-shaped with a characteristic cusp at the base. The dental formula is variable.

● 4. There is no zygomatic arch.

5. There is no auditory bulla and the tympanic bone is ring-shaped and not fused to the skull.

6. They are world-wide in distribution except for Australasia and most of South America. There are 24 genera and 291 species.

Examples:- Sorex, Crocidura.

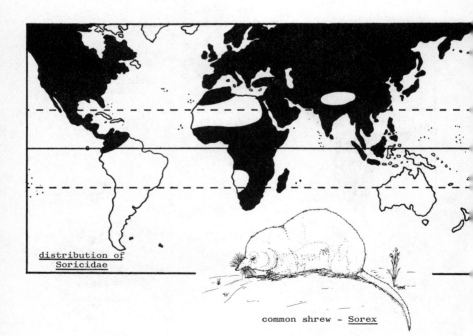

distribution of
Soricidae

common shrew - Sorex

Family Talpidae - Moles

Insectivora in which:-

1. The eyes are small and sometimes covered by the skin. The external ears are usually absent. There is a short tail and the fur is short and thick.

● 2. The fore-limbs and hands are highly modified for digging, the five digits forming a broad spade.

● 3. The upper molars have a W-shaped pattern. (45) The dental formula is I 2-3/1-3, C 1/0-1, Pm 3-4/3-4, M 3/3.

● 4. The zygoma is complete.

5. The auditory bulla is complete.

6. They are found in Europe, Asia and North America. There are 15 genera and 22 species.

Examples:- _Talpa_, _Desmana_.

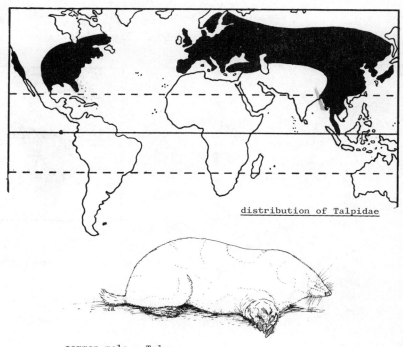

distribution of Talpidae

common mole - _Talpa_

51

Family Macroscelididae - Elephant Shrews

Insectivora in which:-

● 1. The eyes and the ears are large and the tail long.
There is a long, slender, mobile nose.   The fur
is soft (not coarse or spiny).

● 2. The hind legs are long for jumping and the fore
legs relatively short.   The limbs have five digits.

● 3. The upper molars are square in surface view. (45)
The dental formula is I 1-3/3, C 1/1, Pm 4/4,
M 2/2-3.

● 4. The jugal is well developed and the zygoma complete.
It is never fenestrated (as in Tupaiidae).

5. There is a well-developed auditory bulla.

6. They are found in Africa except the Sahara.
There are 5 genera and 28 species.

N.B. Some authorities group the Macroscelididae with
the Tupaiidae as the Menotyphla and place this
suborder either in the Insectivora or with the
Primates.

Examples:- Macroscelides, Elephantulus.

distribution of Macroscelididae

the elephant shrew
Macroscelides

Order Dermoptera, Family Cynocephalidae - Flying lemurs

Eutheria in which:-

1. They are small to medium-sized mammals of arboreal habit, gliding from tree to tree and feeding on leaves, fruit and seeds.

● 2. The limbs support a muscular fold or patagium which extends laterally from the body via the hand and foot from the neck to the tip of the tail. The digits are flattened and the soles of the hands and feet form sucking discs for adhesion and grasping.

● 3. The teeth have sharp cusps. The lower incisors are directed forward (procumbent) and comb-like.(80) The upper incisors are situated at the side of the face leaving the front of the jaw toothless. The canines are like the premolars. The dental formula is I 2/3, C 1/1, Pm 2/2, M 3/3.

4. The skull is broad and flat and the bones fuse early in life so that few sutures remain. The palate has a thickened posterior margin.

5. There is an incomplete post-orbital bar.

6. The lachrymal does not extend on to the face.

7. There is no paroccipital process.

8. They are found only in southeastern Asia. There is 1 genus and 2 species.

Example:- Cynocephalus.

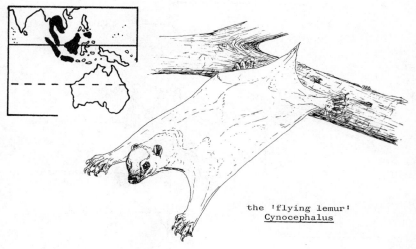

the 'flying lemur'
Cynocephalus

Order Chiroptera - Bats

Eutheria in which:-

● 1. They are very small to medium-sized flying mammals feeding on fruit, nectar, insects or blood according to their group.

● 2. A wing-membrane, or patagium, supported by elongated fingers of the fore-limbs extends from the sides of the body and the legs. The knee is directed backward due to rotation of the hind-limb.

3. There are never more than two pairs of upper incisors. The cheek teeth differ in the two suborders.

4. The skull tends to be elongated, particularly in the Megachiroptera, and the bones to fuse so that few sutures remain.

5. There is rarely a complete post-orbital bar.

6. The lachrymal extends on to the face.

7. There is a small paroccipital process.

<u>ORDER</u>   <u>SUBORDER</u>

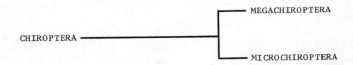

CHIROPTERA ──────────────┬───── MEGACHIROPTERA

                         └───── MICROCHIROPTERA

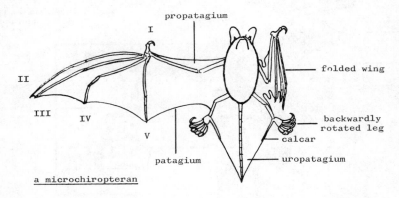

a microchiropteran

54

Suborder Megachiroptera, Family Pteropidae - Fruit bats and
flying foxes

Chiroptera in which:-

- 1. The molars are flat or with a longitudinal groove on
  the surface.   The dental formula is I 1-2/0-2,
  C 1/1, Pm 3/3, M 1-2/2-3.
- 2. The external ear or pinna forms a complete ring
  at the base and is simple.
- 3. The second finger has three joints and a claw.
- 4. The tail, when present, is short and is not
  included in the uropatagium, but lies below it.
  5. They do not echolocate.
  6. They are restricted to the Old World tropics and
  subtropics.   There are 38 genera and 154 species.

Examples:- Pteropus, Eidolon.

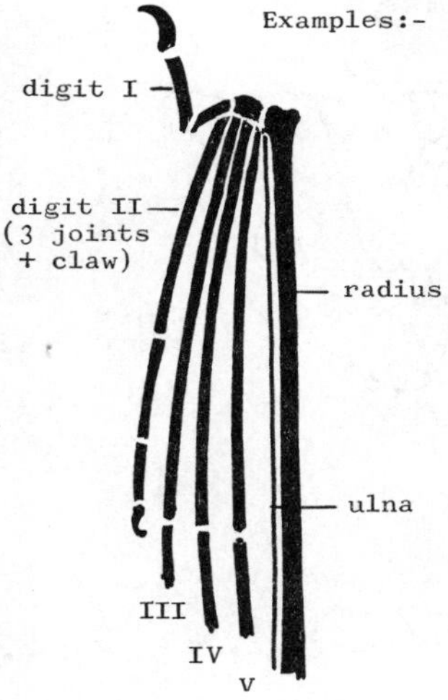

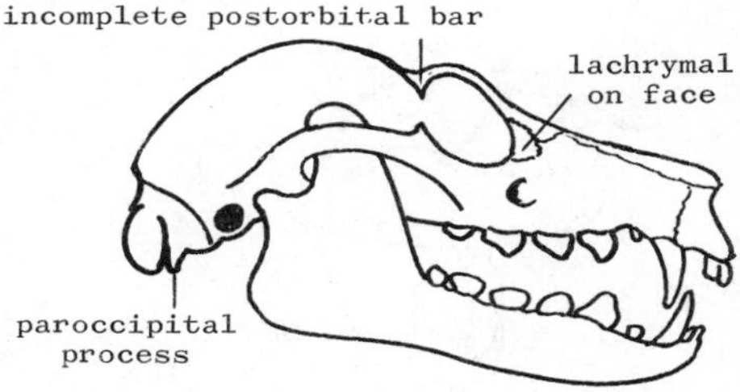

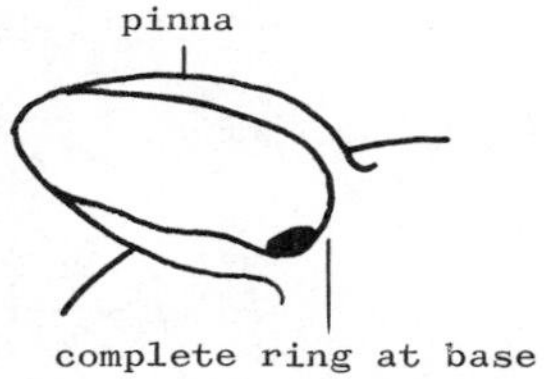

incomplete postorbital bar

digit I

lachrymal
on face

digit II
(3 joints
+ claw)

radius

paroccipital
process

ulna

III

IV

V

pinna

complete ring at base

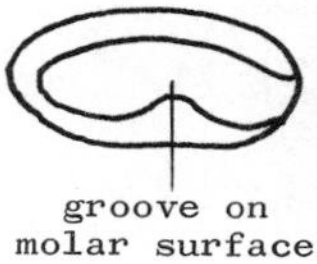

groove on
molar surface

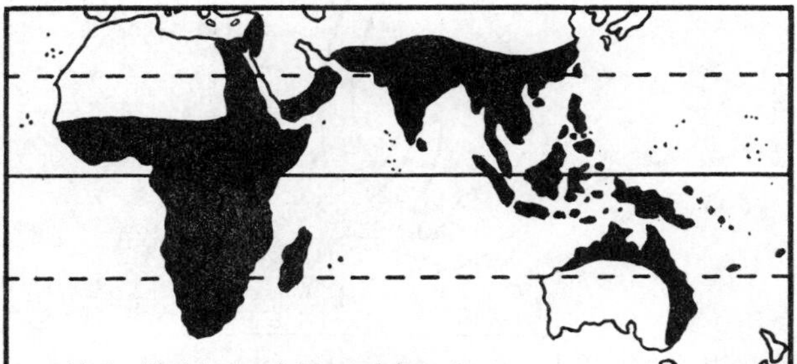

distribution of Megachiroptera

55

Suborder Microchiroptera - Insect-eating bats and vampires

Chiroptera in which:-

- 1. The molars have sharp cusps. The dental formula differs in the various families.
- 2. The external ear or pinna does not form a complete ring at the base and is often complicated.
- 3. The second finger has one or two joints and does not have a claw. (54)
- 4. The tail is usually long and is either included in the patagium or lies above it. (54)
  5. High frequency sounds are produced which, when reflected back to the ear, enable the bat to locate objects and insect prey (echolocation).

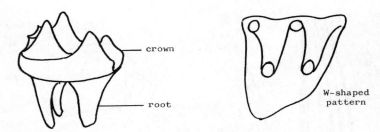

molar with sharp cusps

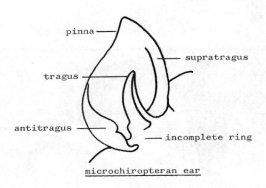

microchiropteran ear

SUBORDER

FAMILY

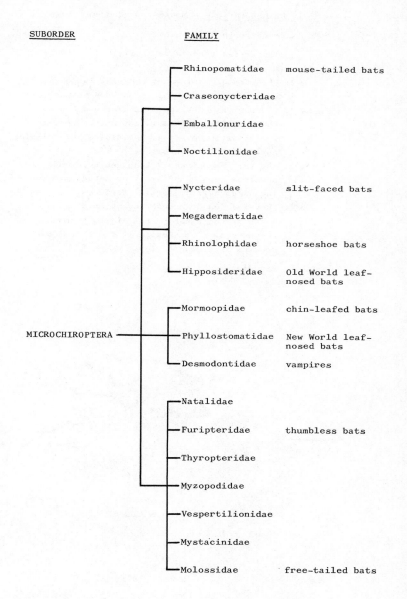

Rhinopomatidae — mouse-tailed bats

Craseonycteridae

Emballonuridae

Noctilionidae

Nycteridae — slit-faced bats

Megadermatidae

Rhinolophidae — horseshoe bats

Hipposideridae — Old World leaf-nosed bats

Mormoopidae — chin-leafed bats

Phyllostomatidae — New World leaf-nosed bats

Desmodontidae — vampires

MICROCHIROPTERA

Natalidae

Furipteridae — thumbless bats

Thyropteridae

Myzopodidae

Vespertilionidae

Mystacinidae

Molossidae — free-tailed bats

Family Rhinopomatidae - Mouse-tailed desert bats

Microchiroptera in which:-

1. The dental formula is I 1/2, C 1/1, Pm 1/2, M 3/3.

● 2. The ears are joined by a mesial lappet and have a clearly visible truncated tragus. (56)

● 3. The nose has a naked, muscular pad around the nostrils.   This is not a true nose-leaf for acoustic emission, but closes the valvular nostrils between breaths.

● 4. The tail is extremely long and thin and projects beyond the small uropatagium.

5. The wings have no special features.

6. They are found in deserts and arid lands north of the equator from Africa through southern Asia to Sumatra.   There are four species in a single genus.

Example:- Rhinopoma.

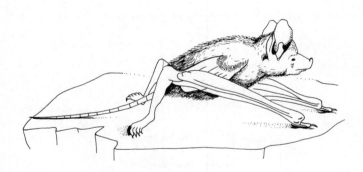

mouse-tailed bat - Rhinopoma

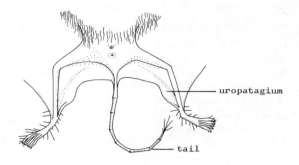

uropatagium

tail

mouse-tailed bat
Rhinopoma

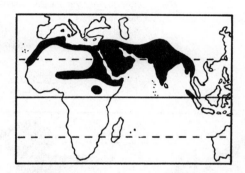

Family Craseonycteridae

Microchiroptera in which:-

1. The dental formula is I 1/2, C 1/1, Pm 1/2, M 3/3.
2. The ears are long and separate and have a clearly visible, tapered tragus. (56)
3. The nose is simple with a narial pad, but open, non-valvular nostrils.
● 4. There is no tail. The uropatagium is of moderate size without a calcar. (54)
● 5. The wing has the second finger partly free in the leading edge.
6. These minute bats are known only from Thailand and were discovered in 1973. There is a single species.

Example:- Craseonycteris.

59

Family Emballonuridae

Microchiroptera in which:-

1. The dental formula is I 1-2/2-3, C 1/1,
   Pm 2/2, M 3/3.
2. The ears are separate and have a small tragus. (56)
3. The nose is simple and often pointed.
● 4. The tail is small and projects from a loose
   sheath onto the dorsal side of the well-
   developed uropatagium.
● 5. The wing often has a glandular sac in the
   propatagium in front of the arm.   The
   elongated 3rd digit folds twice (N-shaped)
   when at rest.
6. These bats often roost in the open or in
   well-lit cave entrances throughout the tropics.
   There are about 40 species in 13 genera.

Examples:- Emballonura, Taphozous, Diclidurus,
            Saccopteryx.

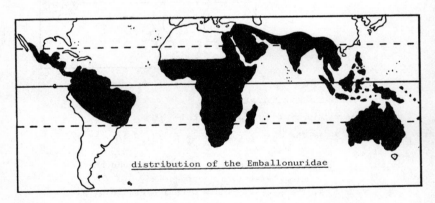

distribution of the Emballonuridae

sac-winged bat - Saccopteryx

Family Noctilionidae

Microchiroptera in which:-

1. The dental formula is I 2/1, C 1/1, Pm 1/2, M 3/3.

2. The ears are large, funnel-shaped and separate and have a very small tragus. (56)

● 3. The nose is simple with deep jowls and a 'hare-lip'.

● 4. The tail perforates the large uropatagium dorsally.

● 5. The wings are long and narrow. The legs are long with enormous feet and powerful claws.

6. They are found in central America, tropical South America and the West Indies. There are two species in a single genus (one feeds mainly on fish and the other is insectivorous).

Example:- Noctilio.

hare-lipped bat - Noctilio

Family Nycteridae - Slit-faced bats

Microchiroptera in which:-

1. The dental formula is I 2/3, C 1/1, Pm 1/2, M 3/3.

● 2. The ears are extremely large (not 'ribbed' as in many other big-eared bats) and separate. The tragus is minute and usually hidden in the long, fine fur.(56)

● 3. The nose has a unique nose-leaf of fleshy pads bounding the nostrils which are in a vertical, facial slit.

● 4. The tail is long and extends to the edge of the very large uropatagium where it ends in a unique T-shaped vertebra and supports the membrane between the long calcars.

5. The wings are broad with a noticeably large area.

6. These minute-eyed bats often hunt non-flying insects and spiders.   They are found in Africa, Arabia, Palestine, Corfu and in the Malay Peninsula to Borneo and Java.   There are about 10 species in a single genus.

Example:-   Nycteris.

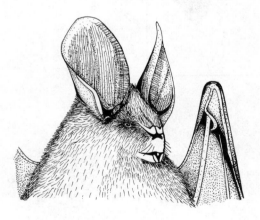

slit-faced bat - Nycteris

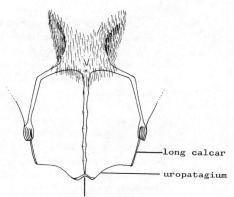

long calcar

uropatagium

terminal T-shaped vertebra

slit-faced bat - <u>Nycteris</u>

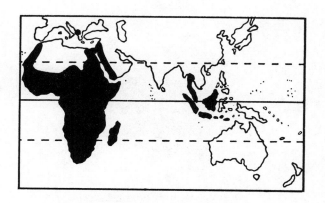

<u>distribution of the Nycteridae</u>

Family Megadermatidae

Microchiroptera in which:-

1. The dental formula is I 0/2, C 1/1, Pm 1-2/2, M 3/3.

● 2. The ears are extremely large and extensively joined.   The tragus is prominent or enormous.(56)

● 3. The nose has a well-developed nose-leaf, especially dorsally, with a central fold that extends laterally and usually conceals the nostrils.

● 4. There is no tail, but a large uropatagium supported by long calcars.(54)

5. The wings have no special features and are not wrapped around the body when the bat roosts by hanging free.

6. These large-eyed bats are widespread in the Old World tropics.   There are 5 species in 4 genera.

Examples:- Megaderma, Lavia.

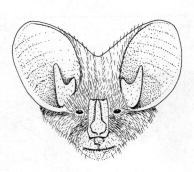

false vampire - Megaderma

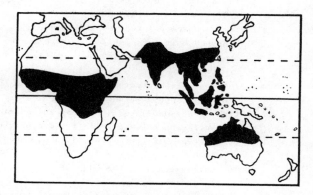

distribution of the Megadermatidae

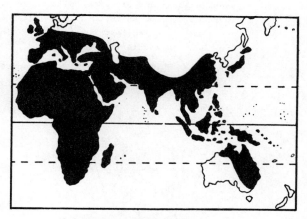

distribution of the Rhinolophidae
including Hipposideridae

Family Rhinolophidae - Horseshoe bats

Microchiroptera in which:-

1. The dental formula is I 1/2, C 1/1, Pm 2/3, M 3/3.

● 2. The ears are pointed, separate and independently very mobile. There is no tragus, but a well-developed antitragus is present. (56)

● 3. The nose has a prominent nose-leaf with the nostrils exposed at the centre of a disc (the 'horseshoe') surmounted by an anteriorly projecting spur (the 'sella') and a pointed, sculptured leaf (the 'lancet') above the eye-line.

4. The tail extends to the edge of the well-developed uropatagium.

● 5. The wings have no special features, but are wrapped around the body with the tail and uropatagium folded dorsally over the back when the bat roosts by hanging free.

6. These insectivorous bats often hang from a vantage point and fly after passing prey. They are widespread in the Old World, but are mainly tropical and subtropical. There are about 50 species in two genera.

Example:- <u>Rhinolophus</u>. (see page 65)

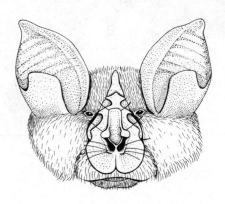

horseshoe bat - <u>Rhinolophus</u>

Family Hipposideridae - Old World leaf-nosed bats

Microchiroptera in which:-

1. The dental formula is I 1/2, C 1/1, Pm 1-2/2,
   M 3/3.

● 2. The ears are pointed, separate or joined, and
   independently very mobile.   There is no
   tragus, but a well-developed antitragus is
   present. (56)

● 3. The nose has a prominent nose-leaf with the
   nostrils exposed at the centre of a disc.
   There is no sella or lancet (see Rhinolophidae),
   but a plain bar dorsally across the eye-line
   or a trident of three processes which may be
   elongated and fingerlike.

4. The tail extends to the edge of the well-
   developed uropatagium.

5. The wings have no special features, but are
   held at the side of the body when the bat
   roosts on walls.

6. These insectivorous bats occur in the Old
   World tropics from Africa to Australia.
   There are about 40 species in 9 genera.

Examples:- Hipposideros, Asellia.

(see page 65)

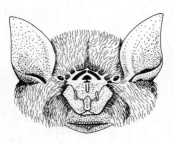

leaf-nosed bat - Hipposideros

Family Mormoopidae - Chin-leafed bats

Microchiroptera in which:-

1. The dental formula is I 2/2, C 1/1, Pm 2/2, M 3/3.

● 2. The ears are well separated and cover the sides of the head to below the eyes. The tragus is small and sometimes inconspicuous.(56)

● 3. The nose has no nose-leaf, but the nostrils are incorporated in a fleshy pad extending onto both lips as an oral 'trumpet' through which ultrasounds are emitted. In Mormoops there are foliar outgrowths on the chin.

4. The tail emerges dorsally from the uropatagium.

● 5. The wings may be joined dorsally over the back ('naked backed').

6. These insectivorous bats are restricted to tropical South America. There are 5 species in two genera.

Examples:- Mormoops, Pteronotus.

distribution of the Mormoopidae

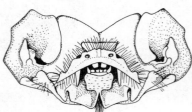

chin-leafed bat - Mormoops

Family Phyllostomatidae - New World leaf-nosed bats

Microchiroptera in which:-

1. The dental formula is variable between species with 26 - 34 teeth.
- 2. The ears are generally well separated and mobile.   The tragus is visible and may be large. (56)
- 3. The nose has a nose-leaf which is generally simple and spear-shaped, but may be hypertrophied or rarely complex.   The nostrils are visible.
- 4. The tail perforates the uropatagium dorsally, but both tail and uropatagium are variable in development and are sometimes absent.
5. The wings have no special features.
6. These bats may take insects or vertebrate prey, feed on fruit or nectar or be omnivorous. They occur from the southwestern United States to north Argentine and in the West Indies. There are 124 species in 48 genera.

Examples:-   Phyllostomus, Glossophaga, Carollia, Sturnira, Artibeus, Phyllonycteris.

distribution of
the Phyllostomatidae

spear-nosed bat - Phyllostomus

Family Desmodontidae - Vampires

Microchiroptera in which:-

1. The dental formula is I 1-2/2, C 1/1,
   Pm 1-2/2-3, M 0-2/0-2.   The incisors are
   uniquely well developed for piercing.
2. The ears are well separated with a clear
   tragus. (56)
● 3. The nose has a nose-leaf which is a fleshy
   pad around the nostrils.
● 4. The tail is absent and the uropatagium
   vestigial.
● 5. The wings have no special features except
   that the 'arms' (and the legs) are strong so
   that vampires can stand with their bodies
   high off the ground and can run or jump with
   great agility.
6. These blood-'sucking' bats feed by cutting
   the skin and 'lapping' the blood with a
   piston-like movement along the axis of the
   tongue.   They occur from north Mexico to
   north Argentine.   There are three species
   in three genera.

Examples:- <u>Desmodus</u>, <u>Diaemus</u>, <u>Diphylla</u>.

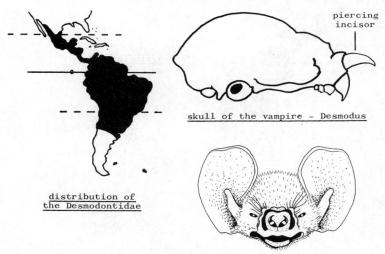

distribution of
the Desmodontidae

piercing
incisor

skull of the vampire - Desmodus

vampire bat - <u>Diphylla</u>

Family Natalidae

Microchiroptera in which:-

1. The dental formula is I 2/3, C 1/1, Pm 3/3, M 3/3.

● 2. The ears are simple, separate, very large and almost surround the minute eyes. The tragus is small but distinct. (56)

● 3. The nose is simple without a nose-leaf, but with lateral hair tufts often forming 'moustaches'.

● 4. The tail extends to the edge of the uropatagium which is extremely large and is supported by the slender, very long legs, the tail and the calcars. (54)

5. The wing membrane looks too big to drive such a tiny body. This bat is nevertheless very agile in flight.

6. These insectivorous bats occur from Baja California through tropical America to Venezuela and Brazil and also in the West Indies. There are 6 species in a single genus.

Example:- Natalus.

long-legged bat - Natalus

Family Furipteridae - Thumbless bats

Microchiroptera in which:-

1. The dental formula is I 2/3, C 1/1, Pm 2/3, M 3/3.
2. The ears are simple, separate, very large and almost surrounding the minute eyes. The tragus is small but distinct.(56)
3. The nose is simple without a nose-leaf.
● 4. The tail lies entirely within the uropatagium, but does not reach the edge.
● 5. The wings have no special features except that the 1st digit (pollex or thumb) is so rudimentary that it appears to be absent.
6. These bats are probably insectivorous and occur in tropical South America and Trinidad. There are two species in separate genera.

Examples:- Furipterus, Amorphochilus.

distribution of the Furipteridae

Family Thyropteridae

Microchiroptera in which:-

1. The dental formula is I 2/3, C 1/1, Pm 3/3, M 3/3.

2. The ears are simple, usually separate and immobile with a pronounced tragus. (56)

3. The nose is simple.

● 4. The tail extends slightly beyond the tip of the well-developed uropatagium.

● 5. The wings have no special features except that there are stalked, adhesive suckers on the wrists, and also on the ankles, which are used for roosting in smooth, rolled banana leaves before they open.

● 6. They are the only sucker-footed bats in the New World.   They occur in tropical America from southern Mexico to Brazil and Peru and in Trinidad.   There are two species in a single genus.

Example:-   Thyroptera.

wrist sucker of Thyroptera

Family Myzopodidae

Microchiroptera in which:-

1. The dental formula is I 2/3, C 1/1, Pm 3/3, M 3/3.

2. The ears are large, simple and separate.

● There is a clear tragus and a unique 'mushroom-shaped' process from the antitragus.

3. The nose is simple.

● 4. The tail projects beyond the edge of the uropatagium.

● 5. The wings have no special features except that there are sessile adhesive suckers on the wrists and also on the feet. The 1st digit (pollex or thumb) is reduced and almost vestigial.

● 6. This bat is found only Madagascar and is extremely rare. There is a single species.

Example:- <u>Myzopoda</u>.

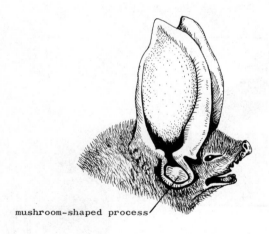

mushroom-shaped process

golden bat - <u>Myzopoda</u>

Family Vespertilionidae

Microchiroptera in which:-

1. The dental formula is I 1-2/2-3, C 1/1,
   Pm 1-3/2-3, M 3/3.
2. The ears are simple, usually separate and
   immobile, with a pronounced sometimes large
   tragus. (56)
3. The nose is simple, although narial pads may
   be present (long-eared bats).
4. The tail extends to the tip of the well-
   developed uropatagium.
5. The wings have no special features except in
   Tylonycteris which has sessile adhesive pads
   on the wrists and feet.   This bat is greatly
   flattened and roosts inside cracked bamboo
   stems in India and the Far East.
6. They are world-wide in distribution, except
   in the extreme north and the Antarctic.
   They are typically insectivores although
   some catch fish.   There are 280 species in
   40 genera.

   Examples:- Vespertilio, Myotis, Pipistrellus,
   Miniopterus, Murina, Kerivoula,
   Nyctophilus, Tomopeas, Tylonycteris.

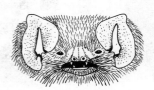

painted bat
Kerivoula

big-eared bat
Nyctophilus

Family Mystacinidae

Microchiroptera in which:-

1. The dental formula is I 1/1, C 1/1, Pm 2/2, M 3/3.

2. The ears are simple, long and well separated with a well developed tragus. (56)

● 3. The nose is simple, but the nostrils are tubular and directed dorsally.

● 4. The tail is thick at the base and perforates the uropatagium dorsally.

● 5. The wings are tucked into thick leathery pockets at the sides of the body when the bat is at rest.   It is very agile on the ground and may feed 'on foot'.   The fur is remarkably thick and all claws bear talons.

● 6. This bat is restricted to New Zealand.   It feeds on insects, fruit, flowers, pollen and possibly carrion.   There is a single species.

Example:- <u>Mystacina</u>.

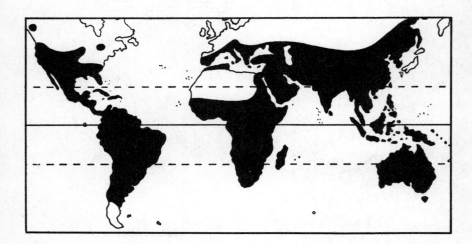

<u>distribution of the Molossidae</u>

<u>Family Molossidae</u> - Free-tailed bats

Microchiroptera in which:-

1. The dental formula is I 1/1-3, C 1/1,
   Pm 1-3/2-3, M 3/3.

● 2. The ears are deeply folded and heavy in
   appearance and may be separate, but are often
   joined by a mesial lappet.   They are lateral
   in position and immobile.

● 3. The nose is simple without a nose-leaf and
   generally overhanging the jaw with wrinkled
   jowls.

● 4. The tail is stout and projects well beyond
   the edge of the uropatagium at rest, although
   the membrane forms a loose sheath and can
   slide freely to the tip of the tail during
   flight.

● 5. The wings are narrow and pointed and often
   very long.

6. These insectivorous bats often form enormous
   colonies in caves and roofs and are world-
   wide in tropical to subtemperate regions.
   There are about 80 species in 12 genera.

Examples:- <u>Molossus</u>, <u>Cheiromeles</u>, <u>Tadarida</u>,
   <u>Eumops</u>.

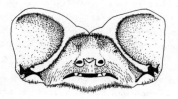

mastiff bat - <u>Eumops</u>

4   The Primates

The Primates are an extensive group of great importance
if only because they include Man. Many of the mammalian orders
contain very few types, at least so far as living mammals are
concerned or they contain a large number at about the same
level of evolutionary development. The Primates on the other
hand, span a wide range of animals from the insectivores to
the higher apes and Man and are exceptional in this respect.
At the lower end of the scale the distinction between the
Primates and the insectivores is a matter for conjecture.
Some authorities count the tree shrews (Tupaiidae) as Primates,
a view which is taken here, while others place them equally
firmly with the elephant shrews in the insectivores. At the
upper end of the scale it might be argued that although from
the anatomical standpoint Man is clearly a Primate with the
apes as his nearest living relatives, from a social view Man
stands so far apart from the rest of animals by virtue of
intellect and group capability as to represent a new class
of vertebrates.

Particular characters diagnostic of the Primates as a
whole are few. On the other hand progression toward an
unmistakable higher primate condition as shown by the grasping
hand, forwardly directed eyes, emotive facial expression and a
tendency to bipedal gait is so clear that there is never much
doubt even to the layman that a monkey is a monkey even if it
is a lemur!

In putting the matching characters together in a form that
makes sense, it sometimes becomes obvious that one grouping or
hierarchy is better on grounds of directness and simplicity
than another often more commonly accepted. The division of
the Primates into Prosimii, Tarsii and Simii brings out these
contrasting characters, but is not found in quite this form
in other classifications.

THE PRIMATES

Order Primates

Eutheria in which:-

1. They are small to large terrestrial or arboreal mammals with plantigrade gait and typically omnivorous diet.(45)
● 2. The limbs are generalised and freely movable with grasping mobile digits in which claws tend to be replaced by flattened nails and enlarged sensitive pads are present on the terminal joints.
3. The teeth are unspecialised in most forms. The incisors may be reduced or enlarged. The premolars and molars have separate, rounded cusps (bunodont) in advanced primates.
4. The skull has a relatively large rounded braincase and a reduced olfactory region particularly in advanced forms. The orbits tend to be directed forward. (87)
● 5. There is a post-orbital bar or plate separating the orbit from the temporal fossa. (80, 87, 92)
6. The lachrymal may or may not extend on to the face.
7. There is no paroccipital process.

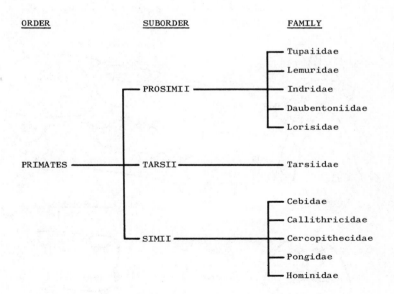

ORDER	SUBORDER	FAMILY
		Tupaiidae
		Lemuridae
	PROSIMII	Indridae
		Daubentoniidae
		Lorisidae
PRIMATES	TARSII	Tarsiidae
		Cebidae
		Callithricidae
	SIMII	Cercopithecidae
		Pongidae
		Hominidae

79

THE PRIMATES

Suborder Prosimii

Primates in which:-

● 1. The face is covered with hair except for the tip of the nose (muzzle or rhinarium) which is hairless and kept moist by glands. The rhinarium extends down onto the upper lip which is divided by a vertical cleft into right and left halves.

● 2. The nostrils have a lateral cleft.

3. The upper lip is held down by a ligament and cannot be protruded. The face in general is also relatively immobile and does not express emotive responses.

● 4. The orbits face antero-laterally and are not separated from the temporal fossa by a bony shelf.

5. The lachrymal bone extends onto the face.

6. The auditory bulla is well developed and there is no bony, tubular, external auditory meatus.

● 7. The lower incisors and canine are directed forward (procumbent) and more or less comb-like except in Daubentonia.

● 8. In the foot the 1st digit always bears a nail (except Tupaiidae) and the 2nd digit always a sharp claw. The 1st digits (thumb and great toe) are opposable except in Tupaiidae.

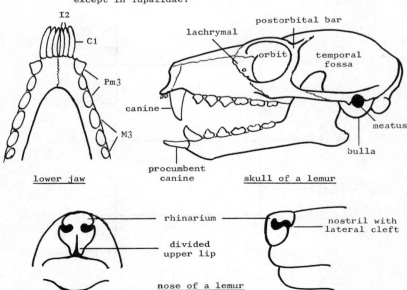

lower jaw

skull of a lemur

nose of a lemur

Family Tupaiidae - Tree shrews

    Prosimii in which:-

     1. The dental formula is I 2/3, C 1/1, Pm 3/3, M 3/3.

● 2. The ears are small and relatively hairless

● 3. The digits bear claws and the 1st on the hand
       and the foot (thumb or pollex and great toe or
       hallux) are not opposable.

     4. They are found in southeast Asia from India to
       the Philippines. There are 5 genera and 15 species.

N.B. The zygoma has a characteristic elongated
     perforation (see Insectivora, Macroscelididae).

    Examples:- Tupaia, Ptilocercus.

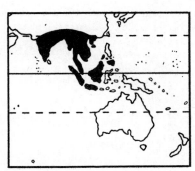

distribution of the Tupaiidae

tree shrew - Tupaia

Family Lemuridae - Lemurs

Prosimii in which:-

● 1. The upper incisors are reduced or absent. The dental formula is I 0-2/2, C 1/1, Pm 3/3, M 3/3.
● 2. The ears are large and covered with hair.
3. The digits of the foot are free (not webbed).
4. They are restricted to Madagascar. There are 5 genera and 15 species.

Examples:- Lemur, Cheirogaleus.

Family Indridae - Indri lemurs

Prosimii in which:-

● 1. The upper incisors tend to be enlarged. The dental formula is I 2/2, C 1/0, Pm 2/2, M 3/3.
● 2. The ears are small and covered with hair.
3. The digits of the foot are webbed.
4. They are restricted to Madagascar. There are 3 genera and 4 species.

Example:- Indri.

Family Daubentoniidae - Aye-ayes

Prosimii in which:-

● 1. The incisors resemble those of the rodents. They have persistent pulps for continuous growth and enamel on the anterior surface only, producing a chisel edge for gnawing and cutting. The canine is often absent and there is a diastema. The dental formula is I 1/1, C 0-1/0, Pm 1/0, M 3/3.
● 2. The ears are large, membranous and movable.
● 3. All digits are free and clawed except the great toe.
4. They are restricted to Madagascar. There is 1 genus and 2 species.

Example:- Daubentonia.

ring-tailed lemur - <u>Lemur</u>

the aye-aye - <u>Daubentonia</u>

indri lemur - <u>Indri</u>

THE PRIMATES

<u>Family Lorisidae</u> - Lorises, Pottos and Galagos or Bushbabies

Prosimii in which:-

1. The dental formula is I 1-2/2, C 1/1, Pm 3/3, M 3/3.
2. The ears are small or moderate in size and membranous.
3. The digits are free and have nails except the 2nd of the foot which has a claw.
4. They are found in Africa, India, southeast Asia and the East Indies.   There are 6 genera and 11 species.

Examples:- <u>Loris</u>, <u>Perodicticus</u>, <u>Galago</u>.

common loris - <u>Loris</u>

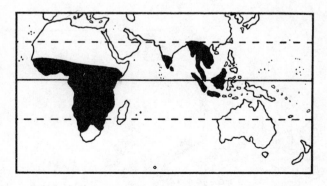

<u>distribution of the Lorisidae</u>

84

THE PRIMATES

Suborder Tarsii, Family Tarsiidae - Tarsiers

Primates in which:-

● 1. The face is covered with short hair except for a
glandular area surrounding each nostril.   The upper
lip is entire, that is not divided into right and
left halves by a vertical cleft. (80)

● 2. A small lateral cleft is present in the nostrils. (80)

3. The upper lip is not protrusible, but the face has
some mobility for emotive expression.

● 4. The orbits are very large, face forward and are
partially separated from the temporal fossa by a bony
shelf joining the post-orbital bar to the cranium. (87)

5. The lachrymal bone extends onto the face. (80)

6. The auditory bulla is well developed and the tympanic
ring forms a tubular, bony, external auditory meatus.

● 7. The lower incisors are vertical.   The dental formula
is I 2/1, C 1/1, Pm 3/3, M 3/3.

● 8. All digits bear nails (except the 2nd and 3rd of the
foot which have claws) and end in adhesive pads.
The 1st digits (thumb and great toe) are opposable.
The hind limbs are long for jumping.

9. They are found in the East Indies from southern
Sumatra to the Philippines.   There is 1 genus
and 3 species.

Example:-   Tarsius.

the spectral tarsier
Tarsius

Suborder Simii

Primates in which:-

● 1. The face is hairless except for particular well-defined regions.   There is no moist muzzle or rhinarium and the upper lip is entire.

● 2. The nostrils are completely surrounded by naked skin and are without a lateral cleft.

3. The upper lip is not held down by ligament and is capable of protrusion.   The face is highly mobile for emotive expression.

● 4. The orbits face forward and are completely separated from the temporal fossa by a bony shelf joining the post-orbital bar to the cranium.

5. The lachrymal bone is within the orbit.

6. The auditory bulla is reduced or absent.   There is a tubular, bony, external auditory meatus in Old World monkeys, but none in the New World monkeys.

● 7. The lower incisors are vertical.

● 8. All the digits bear flattened nails except in the Callithricidae which have a nail on the 1st digit of the foot and pointed claws on the remainder. The 1st digits (thumb and great toe) may be opposable or not or absent.

eyes face
forward

nostrils
without
cleft

entire
upper lip

cheek pouches for food storage

New World (platyrrhine) monkey
(widely separate nostrils
face laterally)

Old World (catarrhine) monkey
(nostrils close and
face downward)

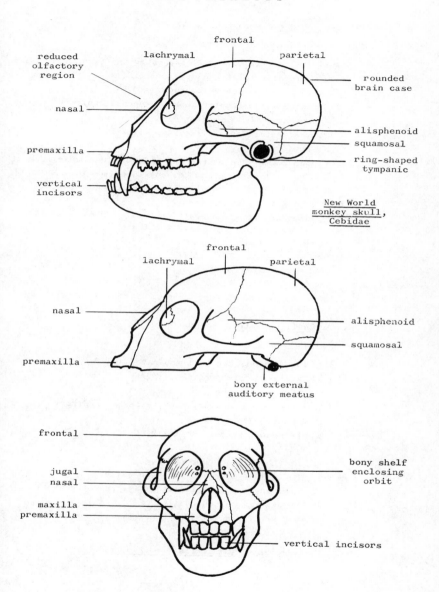

reduced
olfactory
region

lachrymal

frontal

parietal

rounded
brain case

nasal

premaxilla

alisphenoid

squamosal

ring-shaped
tympanic

vertical
incisors

New World
monkey skull,
Cebidae

lachrymal

frontal

parietal

nasal

alisphenoid

squamosal

premaxilla

bony external
auditory meatus

frontal

jugal
nasal

maxilla
premaxilla

bony shelf
enclosing
orbit

vertical incisors

Old World monkey skull, Cercopithecidae

THE PRIMATES

Family Cebidae - New World monkeys

    Simii in which:-

● 1. The nostrils are well separated and face
laterally (platyrrhine). (86)
2. There are no cheek pouches. (86)
3. There is no tubular, bony, external auditory
meatus. (87)
4. The dental formula is usually I 2/2, C 1/1,
Pm 3/3, M 3/3.
5. The thumb when present is slightly opposable.
● 6. The tail is long and prehensile or short.
7. There are no ischial callosities on the buttocks.
8. The gait is quadrupedal.
9. They are found in South America.   There are
11 genera and 29 species.

    Examples:- <u>Cebus</u>, <u>Ateles</u>.

Family Callithricidae - Marmosets

    Simii in which:-

● 1. The nostrils are well separated and face
laterally (platyrrhine). (86)
2. There are no cheek pouches. (86)
3. There is no tubular, bony, external auditory
meatus. (87)
4. The dental formula is I 2/2, C 1/1, Pm 3/3,
M 2-3/2-3.
5. The thumb is not opposable.
● 6. The tail is long and not prehensile.
7. There are no ischial callosities on the buttocks.
8. The gait is quadrupedal.
9. They are found in South America.   There are
4 genera and 33 species.

    Examples:- <u>Callithrix</u>, <u>Leontideus</u>.

spider monkey - <u>Ateles</u>

<u>distribution of</u>
<u>Cebidae and Callithricidae</u>

marmoset - <u>Callithrix</u>

THE PRIMATES

Family Cercopithecidae - Old World monkeys

Simii in which:-

- 1. The nostrils are close together and face downward (catarrhine). (86)
- 2. Cheek pouches are present. (86)
- 3. The tympanic forms a tubular external auditory meatus. (87)
- 4. The dental formula is I 2/2, C 1/1, Pm 2/2, M 3/3.
- 5. The thumb when present is opposable.
- 6. The tail when present is not prehensile.
- 7. There are ischial callosities on the buttocks.
- 8. The gait is quadrupedal.
- 9. They are found in Africa, Arabia and from Afghanistan to Japan.   There are 11 genera and 60 species.

Examples:- Cercopithecus, Papio, Colobus.

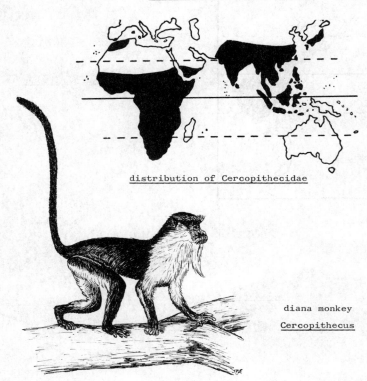

distribution of Cercopithecidae

diana monkey

Cercopithecus

THE PRIMATES

<u>Family Pongidae</u> - Apes

Simii in which:-

- 1. The nostrils are close together and face downward (catarrhine). (86)
- 2. There are no cheek pouches. (86)
- 3. The tympanic forms a tubular external auditory meatus.
- 4. The dental formula is I 2/2, C 1/1, Pm 2/2, M 3/3.
- 5. The thumb when not reduced is opposable and also the great toe.
- 6. There is no tail.
- 7. Ischial callosities on the buttocks are small or absent.
- 8. The gait is semi-bipedal.
- 9. They are found in tropical Africa and southeast Asia.   There are 4 genera and 8 species.

Examples:- <u>Pongo</u>, <u>Gorilla</u>, <u>Chimpansee</u>, <u>Hylobates</u>.

(see page 92)

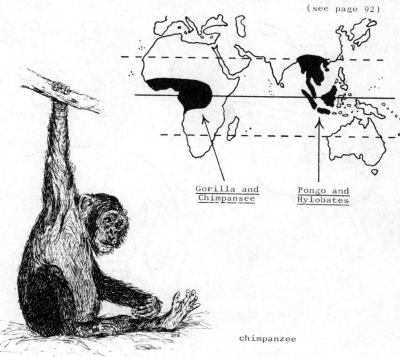

<u>Gorilla and Chimpansee</u>   <u>Pongo and Hylobates</u>

chimpanzee

THE PRIMATES

Family Hominidae - Man

Simii in which:-

● 1. The nostrils are close together and face
downward (catarrhine). (86)

2. There are no cheek pouches. (86)

● 3. The tympanic forms a tubular external auditory
meatus.

4. The dental formula is I 2/2, C 1/1, Pm 2/2, M 3/3.

● 5. The thumb is highly opposable but the great toe
is not opposable.

6. There is no tail.

7. There are no ischial callosities on the buttocks.

● 8. The gait is bipedal.

9. They are world-wide.   There is 1 genus and
1 species.

Example:- Homo.

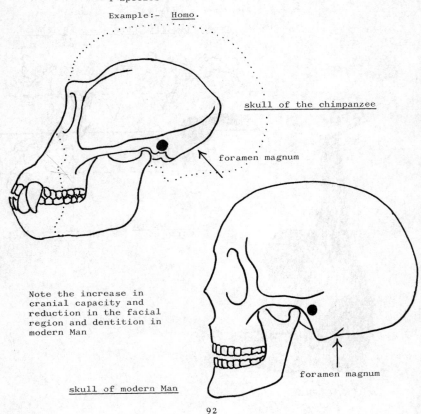

skull of the chimpanzee

foramen magnum

Note the increase in
cranial capacity and
reduction in the facial
region and dentition in
modern Man

foramen magnum

skull of modern Man

92

5 Anteaters and allies

The practice of eating ants has been adopted by a number
of different mammals. To be a successful anteater, ant nests
have to be located by smell, the rock-hard exterior broken open
with powerful claws, and the ants collected on a long, sticky
tongue which can be passed down the galleries of the nest.
Anteaters, therefore, tend to have skulls with a weak lower jaw
and no teeth, a smooth conical shape and a prominent olfactory
region. In some cases mammals that began their evolution as
anteaters, like the sloths and armadillos, changed their diet
and became vegetarians or scavengers before the process of
adaptation to anteating was complete, so that a second set of
features came to overlay and modify the first.

Such powerful convergence as arises from ant eating led
at one time to all anteaters being grouped as the Edentata, but
whereas all anteaters come from very ancient stock they are
evidently from distinct lines and this is now recognised by
placing some of them in different orders.

The order Edentata is reserved for the South American
forms which include not only the anteaters but also the sloths
and armadillos. These animals evidently found their way to
that island continent so early that the general characteristics
of the Eutheria had not by then been fixed. Consequently they
have retained a number of peculiarities not found elsewhere in
Eutheria.

A second order, the Pholidota, contains the pangolin from
Africa and Asia, another very ancient mammal from the earliest
period of Eutherian evolution. Anteaters are inoffensive and
need protection. The pangolin gains this from its covering
of overlapping horny scales and the ability to roll tightly
into a ball.

A N T E A T E R S & A L L I E S

Order Edentata, Suborder Xenarthra

Eutheria in which:-

1. They are small to medium sized terrestrial, burrowing or arboreal mammals with a plantigrade gait. They eat ants and other insects or are vegetarian or scavenge mainly according to their family. (45)
2. The hind foot is typically pentadactyl while the fore-limb has two or three predominant digits with long, sharp recurved claws.
● 3. Incisors and canines are absent. The premolars and ·molars when present are subcylindrical (peg-like) without enamel or roots or with a single root.
4. The skull is variously modified and may be elongated or conical (anteaters), flattened dorso-ventrally (armadillos), or greatly shortened (sloths).
5. There is no postorbital bar.
6. The lachrymal may be large (anteaters) or small (sloths and armadillos) and extends on to the face.
7. There is no paroccipital process.
● 8. There are additional articulations between the lumbar vertebrae to which the name Xenarthra refers.

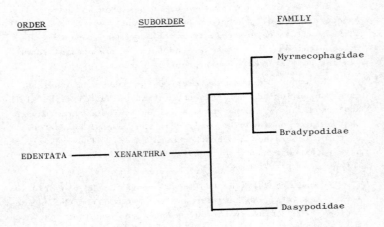

ORDER   SUBORDER   FAMILY

Myrmecophagidae

Bradypodidae

EDENTATA ——— XENARTHRA

Dasypodidae

94

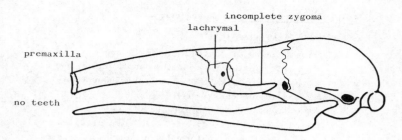

skull of the giant anteater - Myrmecophaga

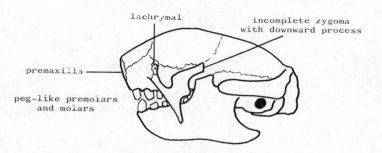

skull of the three-toed sloth - Bradypus

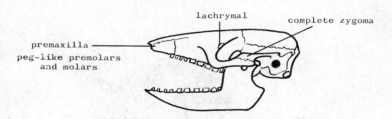

skull of the armadillo - Dasypus

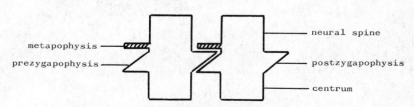

xenarthran vertebrae with additional articulation

A N T E A T E R S & A L L I E S

<u>Family Myrmecophagidae</u> - Anteaters

Edentata in which:-

● 1. The head is elongated, extremely so in
     <u>Myrmecophaga</u>, with a tubular mouth and a long,
     worm-like, sticky tongue and prominent ears.
● 2. There are no teeth. (95)
  3. The limbs are of approximately equal length
     with four or five digits. The 3rd digit of
     the fore-limb is enlarged and has a long curved
     claw for opening ant and termite nests.
  4. There is a long tail which in <u>Tamandua</u> and
     <u>Cyclopes</u> is prehensile.
  5. The fur is thick and there are no bony scutes
     in the skin.
  6. There are seven cervical vertebrae.
  7. <u>Tamandua</u> and <u>Cyclopes</u> are mainly arboreal and
     <u>Myrmecophaga</u> is terrestrial.
  8. They are found in tropical America from southern
     Mexico to northern Argentine. There are
     3 genera and 4 species.

     Examples:- <u>Myrmecophaga</u>, <u>Tamandua</u>, <u>Cyclopes</u>.

collared anteater - <u>Tamandua</u>

ANTEATERS & ALLIES

Family Bradypodidae - Sloths

Edentata in which:-

● 1. The head is short and rounded. The ears are very small and hidden by fur.
● 2. There are five cylindrical teeth in the upper jaw and four in the lower. The grinding surfaces become concave with wear through eating vegetation. The teeth are not replaced (monophyodont). (95)
3. The fore-limbs are considerably longer than the hind-limbs. There are never more than three digits exposed and they have long hook-like claws.
4. The tail is very short.
5. The fur is long and coarse and in the wild has a greenish colour due to algae growing on the hairs. There are no bony scutes in the skin.
● 6. There are usually six cervical vertebrae in Choloepus and nine in Bradypus.
7. They tend to hang upside down in trees.
8. They are found in tropical America from Honduras to northern Argentine. There are 2 genera and 6 species.

Examples:- Bradypus, Choloepus.

distribution of
Myrmecophagidae
and Bradypodidae

three-toed sloth - Bradypus

Family Dasypodidae - Armadillos

Edentata in which:-

- 1. The head is flattened dorso-ventrally and the
    tongue elongated.   The ears are of moderate size.
- 2. There are from seven (Dasypus) to twenty five
    (Priodontes) cylindrical teeth in each jaw.
    They are replaced once (diphyodont).   The
    animals are omnivorous. (95)
  3. The limbs are more or less equal in length.
    The fore-limbs have three to five powerful claws
    for digging.   There are always five digits in
    the hind-limb.
  4. The tail is relatively short and is usually
    encircled by a series of bony rings.
- 5. The dorsal and lateral surfaces of the body are
    protected by transverse bands of bony scutes
    overlaid by horny epidermal plates.   Flexible
    skin between the bands allows some species to roll
    into a ball.   Hair is present between the bands
    and on the ventral surface.
  6. There are seven cervical vertebrae a variable
    number of which are fused together.
  7. They are terrestrial and burrowing.
  8. They are found from Kansas to the Argentine.
    There are 9 genera and 21 species.

    Examples:- Dasypus, Priodontes, Chlamyphorus.

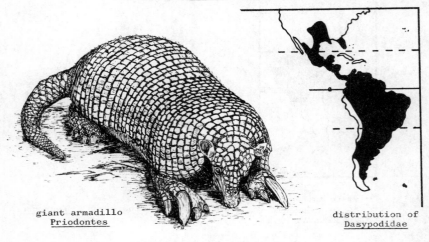

giant armadillo
Priodontes

distribution of
Dasypodidae

ANTEATERS & ALLIES

<u>Order Pholidota</u>, <u>Family Manidae</u> - Pangolins or Scaly anteaters

Eutheria in which:-

●1. They are small to medium sized terrestrial and
   arboreal mammals with a plantigrade gait.   There is
   a very long protrusible, worm-like, sticky tongue used
   for feeding mainly on termites.   The dorsal surface
   of the body is covered with overlapping epidermal
   scales.   The tail is prehensile and the body
   capable of rolling into a ball.

 2. The limbs have strong curved claws used for
   opening termite nests.

●3. There are no teeth.

 4. The skull is conical with an elongated snout and
   a smooth contour.

 5. There is no postorbital bar.

 6. The lachrymal is usually absent.

 7. There is no paroccipital process.

 8. They are found in Africa south of the Sahara and
   southeast Asia.   There is 1 genus and 8 species.

   Example:-   <u>Manis</u>.

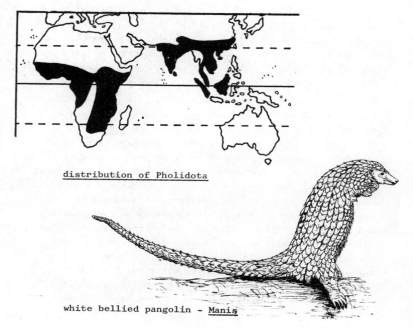

<u>distribution of Pholidota</u>

white bellied pangolin - <u>Manis</u>

99

6 Lagomorphs and rodents

Although they are distantly related if at all, the rabbits,
hares and picas on the one hand and the squirrels, rats and
cavies on the other resemble each other both anatomically and
in their life styles.   All are gnawing vegetarians often
equally at home above or below ground or in the water.
Moreover in the rodents there is a very large number of species,
for they are one of the few mammalian orders that is at or near
its peak and has not yet passed into decline.   More than half
of the living mammals are rodents.

Early classifications lumped the rabbits and hares with
the rodents, but in distinct suborders on the basis of the
upper incisors, two in each half of the jaw (Duplicidentata)
in the rabbits and one (Simplicidentata) in rodents.   But this
has now been discontinued on the grounds that subordinal
separation suggests a phyletic link between the two groups that
is now known not to exist.

The classification of the rodents is a problem, largely
because their relationships are obscured by convergence,
divergence and parallelism.   Nevertheless division into three
main suborders, the squirrels (Sciuromorpha), rats and mice
(Myomorpha) and the cavies (Hystricomorpha) takes care of the
majority of forms, but leaves some such as the Cape Jumping
Hare (Pedetes) and many others in an uncertain position.
Rodent specialists are now tending to move away from this
subordinal division because it represents an over-simplified
and somewhat artificial view of rodent inter-relationships.
But for the student it is probably the most satisfactory
arrangement and has the advantage that the South American
cavies and their allies remain together, although it also
includes the Old World hystricomorphs doubtfully related to
them and probably representing an extreme case of convergence.

Order Lagomorpha

Eutheria in which:-

1. They are small to medium-sized terrestrial or burrowing
   mammals with a digitigrade gait.   The nostrils are
   surrounded by naked skin (rhinarium) which can be
   concealed by the surrounding fur-covered area and the
   nostrils closed.   The tail is a furry tuft or not
   externally visible.   They are vegetarian. (45, 103)

● 2. The fore-limbs have five digits and the hind-limbs
   four or five.   The soles of the feet are covered
   with hair.

● 3. The first upper incisor is large with a persistent
   pulp while the second lies directly behind the first
   (duplicidentate) and is small and peg-like and without
   a cutting edge.   The cheek teeth are tall (hypsodont)
   with persistent pulps and are similar, that is they
   form a molariform row.   There is a wide gap between
   the incisors and the cheek teeth (diastema).   The
   dental formula is I 2/1, C 0/0, Pm 3/2, M 2-3/3. (103)

● 4. In the skull the facial part of the maxillary bones
   is fenestrated. (103)

5. There is no postorbital bar. (103)

6. The lachrymal does not extend on to the face.

7. There is a large paroccipital process. (103)

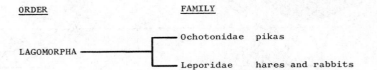

ORDER	FAMILY	
LAGOMORPHA	Ochotonidae	pikas
	Leporidae	hares and rabbits

Family Ochotonidae - Pikas

Lagomorpha in which:-

● 1. The ears are short with a large, valvular
   supratragus and the antitragus more or less
   at right angles to the axis of the pinna. (56)

● 2. The tail is contained in an anogenital
   prominence and is not externally visible.

3. The hind-limbs have four digits and are
   only slightly longer than the fore-limbs
   which have five digits.

● 4. The skull is flat, strongly constricted
   between the orbits and without a supraorbital
   process on the frontal.   The nasals are
   wide anteriorly and the zygoma slender.
   The palate is formed mostly by the palatines.

5. They live in burrows and crevices in
   mountainous regions, forest, scrub and
   steppe from ·southeast Russia to the Arctic
   coast and Japan and, in the New World, in
   Alaska and in California to New Mexico.
   There are 14 species in a single genus.

Example:- <u>Ochotona</u>.

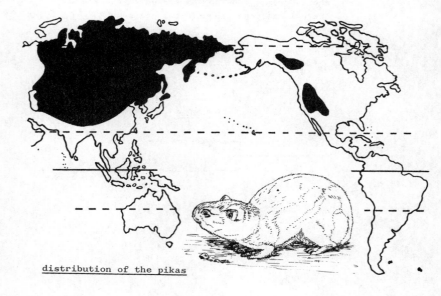

<u>distribution of the pikas</u>

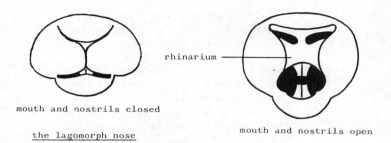

mouth and nostrils closed

the lagomorph nose

rhinarium

mouth and nostrils open

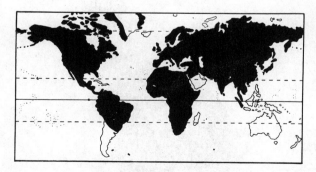

skull of Lepus

distribution of the Leporidae

Family Leporidae - Rabbits and hares

Lagomorpha in which:-

● 1. The ears are long with the supratragus
     reduced to a thin cartilaginous ridge and
     the antitragus in line with the axis of
     the pinna. (56)

● 2. The tail is short and upturned.

  3. The hind-limbs are typically markedly longer
     than the fore-limbs. The first digit is
     reduced on both forefoot and hind foot.

● 4. The skull is arched in profile, moderately
     constricted between the orbits, with a
     prominent supraorbital process on the
     frontal.   The nasals are not wide
     anteriorly and the zygoma is strong and
     vertically expanded.   The palate is formed
     mostly by the maxillae. (103)

  5. They are mostly nocturnal or crepuscular
     animals hiding in burrows or in vegetation
     by day.   They are world-wide in distribution
     except for Australasia, most islands of the
     East Indies, Madagascar, southern Arabia
     and southern South America.   There are
     49 species in 8 genera.

     Examples:- Lepus, Sylvilagus, Oryctolagus.

(see page 103)

hare - Lepus

Order Rodentia

Eutheria ·in which:-

1. They are very small to medium-sized terrestrial, burrowing, arboreal or aquatic mammals with plantigrade or semiplantigrade gait. They are herbivores or scavengers. (45)
2. The limbs usually have five digits with claws. The soles of the feet are not hairy.
●3. There is a single large upper incisor (simplicidentate) with a persistent pulp. The canines and the anterior premolars are missing while the remaining premolars and the molars are alike and form a molariform row with a grinding surface and persistent pulps. There is a wide gap between the incisors and the cheek teeth (diastema). (106, 108)
4. In the skull the facial part of the maxilla is not fenestrated. (107, 108)
5. There is no postorbital bar. (107, 108)
6. The lachrymal bone extends on to the face.
7. There is a paroccipital process of variable dimensions. (107, 108)

ORDER	SUBORDER
	SCIUROMORPHA
RODENTIA	MYOMORPHA
	HYSTRICOMORPHA

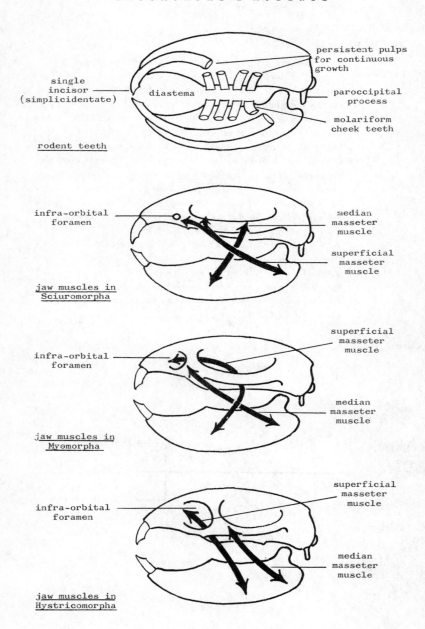

persistent pulps
for continuous
growth

single
incisor
(simplicidentate)

diastema

paroccipital
process

molariform
cheek teeth

rodent teeth

infra-orbital
foramen

median
masseter
muscle

superficial
masseter
muscle

jaw muscles in
Sciuromorpha

superficial
masseter
muscle

infra-orbital
foramen

median
masseter
muscle

jaw muscles in
Myomorpha

superficial
masseter
muscle

infra-orbital
foramen

median
masseter
muscle

jaw muscles in
Hystricomorpha

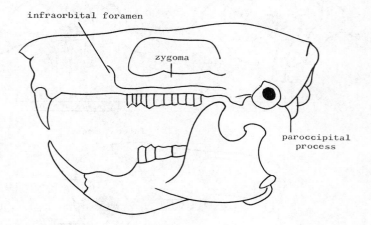

skull of Aplodontia showing
horizontal zygoma

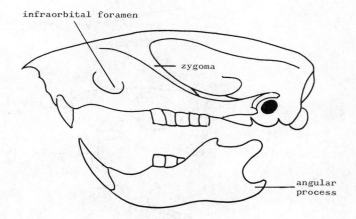

skull of Heteromys showing
the zygoma tilted upward

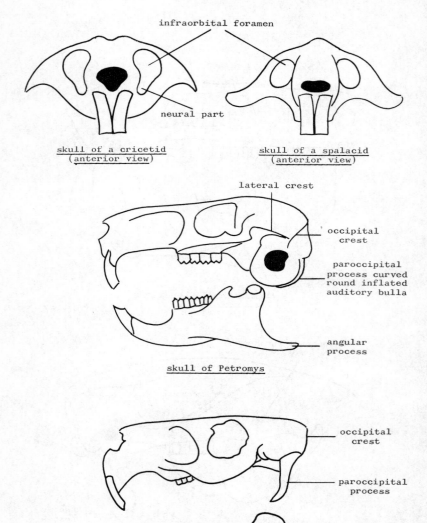

infraorbital foramen

neural part

skull of a cricetid
(anterior view)

skull of a spalacid
(anterior view)

lateral crest

occipital
crest

paroccipital
process curved
round inflated
auditory bulla

angular
process

skull of Petromys

occipital
crest

paroccipital
process

angular
process

skull of Hydrochoerus

L A G O M O R P H S & R O D E N T S

Suborder Sciuromorpha

Rodentia in which:-

1. The dental formula is typically I 1/1, C 0/0,
   Pm 1-2/1, M 3/3.

● 2. The infra-orbital foramen is small and does not
   transmit any part of the masseter muscle except in
   Anomaluridae and Pedetidae.   The superficial masseter
   is inserted partly on the zygomatic arch and partly
   on the face. (106)

SUBORDER   FAMILY

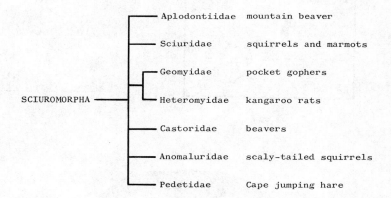

	Aplodontiidae	mountain beaver
	Sciuridae	squirrels and marmots
	Geomyidae	pocket gophers
SCIUROMORPHA	Heteromyidae	kangaroo rats
	Castoridae	beavers
	Anomaluridae	scaly-tailed squirrels
	Pedetidae	Cape jumping hare

Family Aplodontiidae - Mountain beaver

Sciuromorpha in which:-

1. They are burrowing rodents with a heavy,
   thick-set body.   The limbs and tail are
   short, the claws large and the eyes and
   ears small.

● 2. The skull is flat and wide posteriorly with
   a narrow horizontal zygoma below the short,
   broad infraorbital foramen.   The palate
   is broad behind the teeth. (107)

● 3. The cheek teeth have persistent pulps.
   The first upper premolar is minute.   Ridges
   and cusps are almost absent.

4. There is a single species found only in
   mountainous regions in western North America.

   Example:- Aplodontia

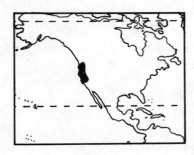

anterior

upper left molar 1

mountain beaver - Aplodontia

Family Sciuridae - Squirrels and marmots

Sciuromorpha in which:-

1. They are arboreal, terrestrial or burrowing
   rodents of variable form and colour.   There
   are four digits on the forefoot and five on
   the hind foot, the fourth being the longest,
   and all have sharp claws.   The tail is often
   ● bushy and always fully haired.   The eyes
   and ears are relatively large.

● 2. The skull is relatively high with a broad
   zygoma tilted upward.   The infraorbital
   canal is short and broad.   The palate is
   broad extending slightly behind or level
   with the teeth with numerous ridges on the
   palatines. (107)

● 3. The cheek teeth are rooted and usually have
   prominent cusps and ridges.   They are
   brachydont or hypsodont.

4. There are about 260 species in 51 genera.
   They are world-wide except for Australasia,
   Madagascar and southern South America.

   Examples:- <u>Sciurus</u>, <u>Funambulus</u>, <u>Callosciurus</u>,
   <u>Xerus</u>, <u>Marmota</u>, <u>Petaurista</u>.

(see page 112)

anterior

<u>upper left molar 1</u>

squirrel - <u>Sciurus</u>

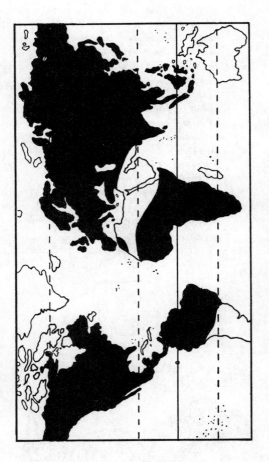

distribution of the Sciuridae

Family Geomyidae - Pocket gophers

Sciuromorpha in which:-

1. They are burrowing rodents. The body is
   thickset. There are five digits with
   powerful claws on the forefoot, the third
   being the longest, and five on the hind
   foot. The tail is short and sparsely haired.
   The eyes and ears are small.

● 2. The skull is flat with a strong zygoma tilted
   upward. The infraorbital canal is long and
   narrow. The palate has a deep pit on each
   side at the level of the third molar and
   behind these there is a strap-shaped
   palatopterygoid plate. (107)

● 3. The cheek teeth have persistent pulps.
   The premolars are 8-shaped and the molars
   mostly circular in surface pattern.

4. They are restricted to North America.
   There are 40 species in 8 genera.

   Examples:- Geomys, Thomomys.

distribution of
the Geomyidae

upper right premolar + molar 1

common pocket gopher - Geomys

113

Family Heteromyidae - Kangaroo rats and spiny mice

Sciuromorpha in which:-

1. They are nocturnal burrowing rodents of
   variable form but generally modified for
   jumping.   The limbs are pentadactyl with
   powerful claws on the forefoot and with the
   hind foot elongated.   The tail is usually
   long and covered with hair.   The eyes are
   large and the ears long.

● 2. The skull is thin and papery, rather
   elongated with a more or less prominent
   projection of the nasals.   The zygoma is
   very slender.   The infraorbital canal is
   long and narrow.   The palate is without
   prominent pits. (107)

● 3. The cheek teeth are typically rooted and
   hypsodont with a two-lobed pattern.

4. They occur in western North America from
   British Columbia through Central America
   to Ecuador, Columbia and Venezuela.   There
   are 75 species in 5 genera.

   Examples:- Heteromys, Dipodomys, Perognathus.

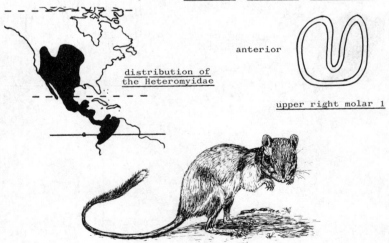

distribution of
the Heteromyidae

anterior

upper right molar 1

kangaroo rat - Dipodomys

Family Castoridae - Beavers

Sciuromorpha in which:-

● 1. They are large aquatic rodents with dense
fur, valvular nostrils and ears and enlarged
webbed hind-feet.   The limbs are pentadactyl
with strong claws, the first and second of
the hind foot being cleft for grooming.
The tail is a horizontal, scaly paddle almost
devoid of hair.   The eyes are small and the
ears short.   Castors, or paired anal scent
glands, are present in both sexes.

● 2. The skull is massive with a broad zygoma
tilted upward.   The infraorbital canal is
long and narrow.   The palate is without
prominent pits. (107)

● 3. The cheek teeth are rooted and hypsodont
with narrow inner and outer folds of enamel.

4. They are found in North America, Europe and
northern Asia.   There are two species in a
single genus.

Example:- <u>Castor</u>.   (see page 116)

anterior

<u>upper left molar 1</u>

beaver - <u>Castor</u>

115

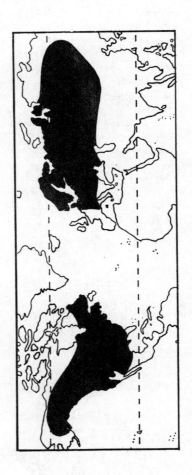

distribution of the Castoridae

Family Anomaluridae - Scaly-tailed squirrels

Sciuromorpha in which:-

1. They are arboreal squirrel-like rodents
   usually brightly coloured and with a
   • patagium for gliding (except Zenkerella)
   supported laterally by an olecranon
   cartilage. They are pentadactyl with
   strong claws. The tail may be very long,
   bushy on the upper surface and always with
   • a double row of raised scales beneath.
   The eyes and ears are relatively large.
• 2. The skull is relatively high with a narrow
   horizontal zygoma beneath a short wide
   infraorbital foramen. The palate is
   extremely narrow. (107)
• 3. The cheek teeth are rooted and brachydont
   each with 4 or 5 cusps.
4. They are restricted to West and central
   Africa. There are 12 species in 4 genera.

   Examples:- Anomalurus, Anomalurops, Idiurus,
              Zenkerella.

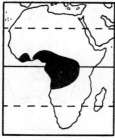

distribution of
the Anomaluridae

anterior

upper left molar 1

scaly-tailed squirrel
Anomalurops

Family Pedetidae - Cape jumping hare

Sciuromorpha in which:-

1. They are relatively large terrestrial and
   burrowing rodents.   They are nocturnal and
   move by hopping and leaping.   There are five
   digits with long claws on the forefoot and
   four with strong flattened claws on the hind
   foot.   The fore-limbs are very short and
   the hind-limbs long for jumping.   The tail
   is long with long hair.   The eyes are large
   ● and the ears long and capable of closure
   while digging by rolling the pinna
   longitudinally.

● 2. The skull is high and massive with a thickened
   horizontal zygoma but not widely spreading.
   The infraorbital canal is short and greatly
   enlarged.   The palate is very short. (107)

● 3. The cheek teeth are brachydont and have
   persistent pulps.   The cusps are simple
   and transversely cleft.

4. They are restricted to central and southern
   Africa.   There are two species in a single
   genus.

   Example:-   Pedetes.

anterior

lower left molar 1

Cape jumping hare - Pedetes

Suborder Myomorpha

Rodentia in which:-

1. The dental formula is typically I 1/1, C 0/0,
Pm 0/0, M 3/3.   The premolars when present are
small and shed early in life.

● 2. The infra-orbital foramen is moderately enlarged and
transmits the median part of the masseter muscle which
is inserted on the face.   The superficial part is
also inserted on the face. (106)

SUBORDER   FAMILY

                         ┌─ Cricetidae   hamsters, voles,
                         │   lemmings and gerbils

                         ├─ Spalacidae   mole rats

                         ├─ Rhizomyidae   bamboo rats

                         └─ Muridae   rats and mice

MYOMORPHA ───   ┌─ Gliridae   dormice

                         ├─ Platacanthomyidae spiny dormice

                         └─ Seleviniidae   dzhalmans

                         ┌─ Zapodidae   jumping mice

                         └─ Dipodidae   jerboas

Family Cricetidae - Hamsters, voles, lemmings, gerbils,
                   and New World rats and mice

Myomorpha in which:-

1. They are primarily terrestrial rodents,
   running and hopping while some burrow and
   others are semi-aquatic, and are
   correspondingly adapted.

2. The skull varies greatly in form.   The
   zygoma is broad and tilted upward.   The
   ● infraorbital foramen has a rounded upper
   portion and a lower (neural) narrower portion.
   In the lower jaw the angular process is not
   distorted outward by the masseter muscle. (108)

3. The cheek teeth vary from brachydont to
   hypsodont.   The occlusal pattern is based
   ● on five cusps arranged in two longitudinal
   rows.

4. They are world-wide except for Australasia
   and Malaysia.   There are about 567 species
   in 97 genera.

      Examples:-   Cricetus, Calomys, Myospalax,
                   Nesomys, Lophiomys, Microtus,
                   Lemmus, Gerbillus.

posterior

upper right molar 2

Norwegian lemming - Lemmus

Family Spalacidae - Mole rats

Myomorpha in which:-

1. They are mole-like rodents living mostly
   underground, but often coming to the surface
   at night.   The fore-limbs are short and
   powerful and the hind-limbs longer and more
   slender;   they are pentadactyl but without
   enlarged claws.   The head is rounded and
 ● flattened and the eyes and ears vestigial.
   There is no externally evident tail.

2. The skull has a high, broad occipital region
 ● sloping forward.   The zygoma is outwardly
   bowed and horizontal thus eliminating the
   lower neural part of the infraorbital foramen.
   In the lower jaw the angular process is
   distorted outward by the masseter muscle. (108)

● 3. The cheek teeth are rooted and semi-hypsodont
   with inner and outer re-entrant folds of
   enamel reduced to islands on the occlusal
   surface as wear proceeds.   The lower incisors
   are very large.

4. They are found in southeastern Europe and
   the eastern Mediterranean region.   There
   are 3 species in a single genus.

   Example:- Spalax.

anterior

upper left molar 1

mole rat - Spalax

Family Rhizomyidae - Bamboo rats

Myomorpha in which:-

1. They are burrowing rodents of mostly
   crepuscular habit, feeding above ground.
   The body is compact and the limbs short,
   pentadactyl and with strong claws. The
   head is broad and flat and the eyes and ears
   relatively small or reduced but functional.
● The tail is short, about one third of the
   body length, without scales, but scantily
   haired.

2. The skull has a high broad occipital region
   slightly tilted forward. The zygoma is
   strong, outwardly bowed and fused anteriorly
   to the rostrum to eliminate the lower neural
   part of the infraorbital foramen. In the
   lower jaw the angular process is not markedly
   distorted outward by the masseter muscle.

● 3. The cheek teeth are rooted and semi-hypsodont
   with inner and outer re-entrant folds of
   enamel reduced to islands on the occlusal
   surface as wear proceeds.

4. They are found in tropical East Africa and
   east Asia.   There are 18 species in 3 genera.

   Examples:- Rhizomys, Cannomys, Tachyoryctes.

anterior

upper left molar 1

bamboo rat - Rhizomys

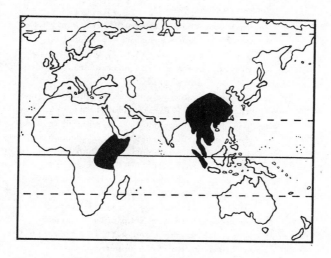

distribution of the Rhizomyidae

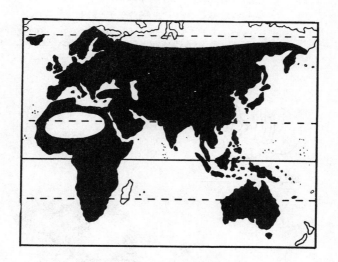

distribution of the Muridae

Family Muridae - Old World rats and mice

Myomorpha in which:-

1. They are typically small terrestrial,
   arboreal, burrowing or semi-aquatic rodents.
   The limbs are pentadactyl but the first digit
   of the fore-limb is rudimentary.   The tail
   is long and scaly.

2. The skull shows no extreme modifications.
   The zygoma is narrow and tilted upward with
   ● the jugal bone typically reduced to a splint.
   The infraorbital canal is not conspicuously
   wider in the upper region than in the lower
   neural part.   In the lower jaw the angular
   process is not markedly distorted outward
   by the masseter muscle.(107)

● 3. The cheek teeth are typically rooted and
   brachydont with a cuspidate or laminate
   occlusal pattern and with tubercles on the
   lingual side of the upper molars.

4. They are widespread in the Old World and,
   by introduction, world-wide.   There are
   about 457 species in 98 genera.

Examples:-   Mus, Rattus, Dendromus, Otomys,
             Phloeomys, Rhynchomys, Hydromys.

(see page 123)

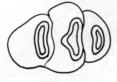

posterior

upper left molar 1

brown rat - Rattus

Family Gliridae - Dormice

Myomorpha in which:-

1. They are small, nocturnal, mainly arboreal,
   squirrel-like rodents.   They are pentadactyl
   but the first digit of the fore-limb is
   rudimentary.   The muzzle tends to be
   elongated and the eyes and ears well developed.
   The tail is bushy.

● 2. The skull is relatively elongated.   The
   zygoma is more or less horizontal with its
   anterior face nearly vertical.   The
   infraorbital canal extends above the median
   level of the orbit.   In the lower jaw the
   angular process is typically inflected and
   in Graphiurus is perforated. (107)

● 3. The cheek teeth are brachydont with a
   characteristic occlusal pattern of transverse
   parallel ridges of enamel.

4. They are found in Europe, Asia Minor, North
   Africa, Africa south of the Sahara and Japan.
   There are 23 species in 7 genera.

Examples:- Glis, Glirulus, Graphiurus.

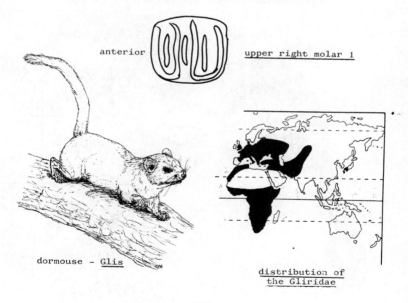

anterior   upper right molar 1

dormouse - Glis

distribution of
the Gliridae

Family Platacanthomyidae - Spiny dormice

Myomorpha in which:-

1. They are little known arboreal mice gnawing
   holes in the tops of trees in rocky ravines.
   The limbs are pentadactyl, but with the first
   digit of the fore-limb reduced to a pad and
   the hind-foot elongated and slender.   The
   eyes are normal (Platacanthomys) or very
   ● reduced (Typhlomys).   The fur is spiny and
   the tail bushy (Platacanthomys) or soft with
   a relatively long tail, scaly and poorly
   haired at the base (Typhlomys).

2. The skull is high with a slender rostrum and
   prominent supraorbital ridges.   The zygoma
   is more or less horizontal and narrow.   The
   infraorbital foramen is large.   In the lower
   jaw the angular is distorted outward by the
   masseter muscle. (107)

● 3. The cheek teeth are subhypsodont with oblique
   ridges of enamel becoming transverse at the
   posterior end of the tooth row.

4. They are found in southern India and southern
   China.   There are two species in separate
   genera.

   Examples:- Platacanthomys, Typhlomys.

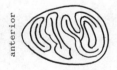

upper left molar 1

spiny dormouse -
Platacanthomys

distribution of the
Platacanthomyidae

126

Family Seleviniidae - Dzhalmans

Myomorpha in which:-

1. They are small, compact, mainly insectivorous
   rodents, living in burrows, with four digits
   on the forefoot and five on the hind.   The
   eyes are normal and the ears large and mobile.
   The tail is long and covered with short hairs.
2. The skull is somewhat rounded with a broad
   occipital region and very large tympanic
   bullae.   The zygoma is weak and tilted upward
   and the infraorbital foramen small.   In the
   lower jaw the angular process is perforated.(107)
3. The cheek teeth are very small, rooted and
   brachydont.   The occlusal pattern is simple
   being concave and smooth.   The upper incisors,
   however, are large.
4. They are found only in the Betpakdala
   Desert in central Asia.   There is a
   single species.

Example:-   Selevinia.

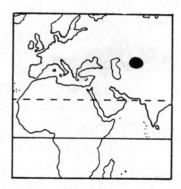

dzhalman - Selevinia

Family Zapodidae - Jumping mice

Myomorpha in which:-

1. They are small or medium sized mouse-like
   running or jumping rodents.   The limbs are
   pentadactyl with the hind-limbs little
   elongated (Sicista) or considerably lengthened
   for jumping (Zapus).   The eyes and ears are
   moderately large.   The tail is long.

2. The skull is relatively high with a simple
   zygoma slightly tilted upward and a large
   infraorbital foramen.   The lower jaw is
   weak with the angular process not distorted
   outward by the masseter muscle.(107)

● 3. The cheek teeth are brachydont (Sicista) or
   semihypsodont (Zapus) with four cusps
   (quadritubercular) and re-entrant folds of
   enamel on the occlusal surface.

4. They are widespread in the Holarctic mainly
   south of the Arctic Circle.   There are 11
   species in 4 genera.

   Examples:- Sicista, Zapus.

distribution of the Zapodidae

posterior

upper left molar 1

American jumping mouse - Zapus

<u>Family Dipodidae</u> - Jerboas

Myomorpha in which:-

1. They are small to moderate-sized jumping,
   usually bipedal, arid plains or desert
   rodents.   They are nocturnal and crepuscular,
   living by day in burrows.   The fore-limb is
   pentadactyl.   The hind-limb is greatly
   elongated with the three central metatarsals
   fused, the lateral metatarsals reduced or
   absent and the lateral hind digits reduced
   and non-functional or absent.   The eyes
   and ears are typically large.   The tail is
   very long and terminally tufted.

2. The skull is rounded with a greatly enlarged
   auditory region and bulla.   The zygoma is
   narrow and weak and the infraorbital foramen
   considerably enlarged.   The lower jaw is
   weak with the angular process not outwardly
   distorted, but frequently perforated.

3. The cheek teeth are hypsodont and rooted
   with a cuspidate, heptamerous occlusal
   pattern.

4. They are found in North Africa and in Asia
   from Asia Minor to China.   There are 27
   species in 10 genera.

   Examples:- <u>Dipus</u>, <u>Cardiocranius</u>, <u>Euchoreutes</u>,
   <u>Allactaga</u>.

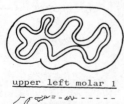

anterior

upper left molar 1

Egyptian jerboa - <u>Dipus</u>

<u>distribution of</u>
<u>the Dipodidae</u>

LAGOMORPHS & RODENTS

Suborder Hystricomorpha - Porcupines and Cavies

Rodentia in which:-

1. The dental formula is typically I 1/1, C 0/0, Pm 1/1, M 3/3.

● 2. The infra-orbital foramen is greatly enlarged and transmits an enlarged median part of the masseter muscle which is inserted on the face. The superficial part retains the more primitive attachment to the zygomatic arch. (106)

SUBORDER   FAMILY

	Hystricidae	Old World porcupines
	Erethizontidae	New World porcupines
	Caviidae	guinea pigs
	Hydrochoeridae	capybaras
	Dinomyidae	pacarana
	Dasyproctidae	agoutis and pacas
	Chinchillidae	chinchillas and vizcachas
	Capromyidae	hutias and coypus
HYSTRICOMORPHA	Octodontidae	hedge rats
	Ctenomyidae	tuco-tucos
	Abrocomidae	rat chinchillas
	Echimyidae	spiny rats
	Thryonomyidae	cutting grass
	Petromyidae	rock rat
	Bathyergidae	mole rats
	Ctenodactylidae	gundis

130

Family Hystricidae - Old World porcupines

Hystricomorpha in which:-

1. They are medium-sized to large ground-living
rodents with the hairs more or less modified
as spines or bristles. The body is thick-set
with short legs, strong for digging and
bearing five digits, but with the first digit
of the fore-limb reduced. The eyes are set
relatively far back on the head and the ears
are rather small. The tail is of moderate
length or short and bears modified spines
or bristles.

2. The skull is massive with large occipital
ridges but small paroccipital processes.
The bones of the nasal region are pneumatised
and enlarged.

3. The cheek teeth are hypsodont and rooted with
re-entrant enamel folds on the occlusal
surface reduced to islands by wear.

4. They are found in Africa, Italy, southern
Asia and the East Indies. There are 15
species in 4 genera.

Examples:- Hystrix, Atherurus.

(see page 132)

anterior

upper left molar 1

African brush-tailed
porcupine - Atherurus

131

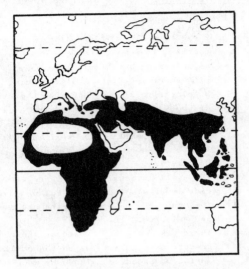

distribution of the Hystricidae

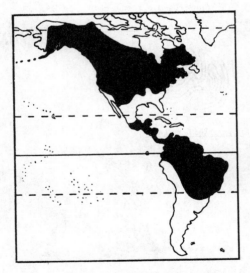

distribution of the Erethizontidae

<u>Family Erethizontidae</u> - New World porcupines

Hystricomorpha in which:-

1. They are medium-sized arboreal rodents with
   some hairs modified as short spines with
   minute barbs.   The body is thickset with
   moderately short legs each with four digits,
   the first being reduced, bearing long curved
   claws.   The eyes and ears are of moderate
   size. ' The tail may be short (<u>Erethizon</u>) or
   long (<u>Chaetomys</u>) and is prehensile curling
   dorsally in <u>Coendou</u>.

2. The skull is relatively high (particularly so
   in <u>Coendou</u>) with prominent occipital ridges
   but small paroccipital processes.

3. The cheek teeth are subhypsodont and rooted
   with large re-entrant enamel folds.

4. They occur widely in North America and
   tropical South America.   There are 8 species
   in 4 genera.

Examples:- <u>Erethizon</u>, <u>Coendou</u>, <u>Chaetomys</u>.

posterior

<u>upper left molar 1</u>

tree porcupine - <u>Coendou</u>

133

Family Caviidae - Guinea pigs

Hystricomorpha in which:-

1. They are small to medium-sized ground-living
   rodents.   The body is sturdy with the legs
   short (Cavia) or long (Dolichotis) and with
   the digits of the forefoot reduced to four
   and of the hind foot to three.   The claws
   vary in form with the genus.   The eyes are
   large and the ears short in Cavia and long
 ● in Dolichotis.   The tail is vestigial.
2. The skull has low occipital ridges and short
 ● paroccipital processes.   The palate is
   very short.
 ● 3. The cheek teeth are hypsodont with persistent
   pulps.   The tooth rows converge anteriorly.
   The occlusal pattern is simple formed by a
   wide infolding of enamel on each side of
   the tooth.
4. They occur widely in South America.   There
   are 12 species in 5 genera.

   Examples:- Cavia, Dolichotis.

posterior

lower left molar 1

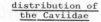

distribution of
the Caviidae

guinea pig - Cavia

134

Family Hydrochoeridae - Capybara

Hystricomorpha in which:-

1. They are the largest of all rodents reaching
   a weight of more than 50 Kg. The body is
   sparsely haired and massive and the legs
   relatively short with four digits on the
   forefoot and three on the hind foot. The
   digits are partially webbed for a semi-
   aquatic life. The eyes are small and the
   ears short and both are high and relatively
   far back on the head, remaining above water
   when the animal is almost submerged. The
   tail is vestigial.

2. The skull is massive with low occipital ridges
   and very long paroccipital processes. (108)

3. The cheek teeth are hypsodont with persistent
   pulps. The occlusal pattern is a series of
   oblique ridges formed by infolding of enamel
   separated by cement. The main part of the
   tooth row is formed by the third molar.

4. They occur in tropical South America east of
   the Andes. There are two species in a
   single genus.

   Example:- Hydrochoerus.

posterior

upper left molar 3

capybara - Hydrochoerus

135

Family Dinomyidae - Pacarana

Hystricomorpha in which:-

1. They are medium-sized to large ground-living
   rodents with white stripes along the back
   and spots on the flanks.   The legs are short
   and the feet broad with four digits on each
   with long powerful claws.   The eyes are
   normal and the ears short.   The tail is
   relatively short and fully haired.

2. The skull is stout with low occipital ridges
   and small paroccipital processes.   The palate
   is constricted anteriorly.

3. The cheek teeth are strongly hypsodont with
   an occlusal pattern of transverse ridges
   deeply separated into plates.

4. They are rare and are found in tropical South
   America.   There is a single species.

Example:- <u>Dinomys</u>.

posterior

<u>upper left molar 1</u>

pacarana - <u>Dinomys</u>

Family Dasyproctidae - Agoutis and pacas

Hystricomorpha in which:-

1. They are medium-sized running rodents with
   long legs and functional digits reduced to
   four in the fore-limb and three in the hind
   • limb, all with hoof-like claws.   The eyes
   are large and the ears of moderate size.
   The tail is short or absent.

2. The skull is relatively high with occipital
   ridges and short paroccipital processes.
   In Cuniculus the zygoma is enormously enlarged.

• 3. The cheek teeth are strongly hypsodont and
   semi-rooted with an occlusal pattern of
   transverse ridges forming islands with wear.

4. They are found in tropical South America.
   There are 11 species in 4 genera.

   Examples:- Dasyprocta, Cuniculus.

anterior

upper left molar 1

agouti - Dasyprocta

distribution of
the Dasyproctidae

Family Chinchillidae - Chinchillas and viscachas

Hystricomorpha in which:-

1. They are medium-sized, ground-living rodents
   sheltering in crevices or burrows.   The hind-
   limbs are relatively long with four digits
   (Chinchilla, Lagidium) or three (Lagostomus)
   and the fore-limbs short with four or five
   digits.   The eyes are large and the ears
   are large or moderate in size and rounded.
   The tail is relatively long and well-haired
   or bushy.
2. The skull tends to be elongated with occipital
   ridges (prominent in Lagostomus) and short
   or long (Lagostomus) paroccipital processes.
   The palate is constricted anteriorly.
● 3. The cheek teeth are hypsodont with persistent
   pulps and an occlusal pattern of transverse
   ridges of enamel not separated by cement.
4. They are restricted to the southern part of
   South America.   There are 6 species in 3
   genera.

   Examples:- Chinchilla, Lagidium, Lagostomus.

posterior

upper left molar 1

short-tailed chincilla - Chinchilla

distribution of
the Chinchillidae

138

Family Capromyidae - Hutias and coypus

Hystricomorpha in which:-

1. They are relatively large ground-living,
   arboreal or semi-aquatic rodents with short
   legs.   The limbs are pentadactyl with strong
   claws but the first digit of the fore-limb
   is reduced or vestigial.   In Myocastor the
   first four digits of the hind-foot are webbed.
   The eyes and ears are small.   The tail is
   reduced, moderate in length or long and
   prehensile (Capromys) and is haired or
   poorly haired.
2. The skull is massive with prominent occipital
   ridges and with moderately long or long
   (Myocastor) paroccipital processes.
● 3. The cheek teeth are hypsodont with persistent
   pulps or are semi-rooted (Myocastor) and an
   occlusal pattern of inner and outer enamel
   folds forming W-shaped ridges which may
   isolate as islands with wear.   The enamel
   folds are filled with cement or not (Myocastor).
4. They are found in the West Indies and
   temperate South America (Myocastor).   There
   are 8 species in 4 genera.

Examples:- Capromys, Geocapromys, Plagiodontia,
Myocastor.

posterior

upper left molar 1

distribution of
the Capromyidae

hutia - Capromys

Family Octodontidae - Octodonts

Hystricomorpha in which:-

1. They are medium-sized ground-living or
   burrowing rodents that shelter in holes,
   crevices and extensive burrows.   The legs are
   short with five digits on the fore-limb and
● four on the hind-limb.   Each hind digit has
   a row of stiff bristles extending beyond the
   claw.   The eyes are moderately large and so are
● the rounded ears.   The tail is long or short
   with hairs progressively longer toward the
   tip forming a trumpet-shaped tuft.
2. The skull has low occipital ridges and very
   short paroccipital processes.
● 3. The cheek teeth are hypsodont with persistent
   pulps and a simple 8-shaped occlusal pattern.
4. They are found on the west coast of South
   America from southern Peru to the Argentine.
   There are 8 species in 5 genera.

Examples:- Octodon, Spalacopus.

distribution of
the Octodontidae

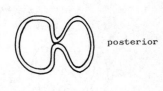

posterior

lower left molar 1

South American bush rat - Octodon

140

Family Ctenomyidae - Tuco-tucos

Hystricomorpha in which:-

- 1. They are medium-sized burrowing rodents with a compact body, large head and short legs.   The soles of the feet are bordered with comb-like bristles.   They are pentadactyl with very strong claws.   The eyes and ears are small.   The tail is short and sparsely covered with short hair.
- 2. The skull is strong and flattened with prominent occipital crests and large paroccipital processes curved beneath the bullae.
- 3. The cheek teeth are hypsodont with persistent pulps and a simple, kidney-shaped occlusal pattern due to the fact that the inner enamel fold is absent.   The last molar is vestigial.
   4. They are widespread in the southern half of South America.   There are 26 species in a single genus.

Example:- Ctenomys.

posterior

upper left molar 1

tuco-tuco - Ctenomys

141

Family Abrocomidae - Rat chinchillas

Hystricomorpha in which:-

1. They are medium-sized rat-like rodents
   living colonially in burrows and crevices
   and capable of climbing trees. The limbs
   are short with four digits on the fore-limb
   and five on the hind, all bearing weak nails.
   ● Stiff hairs project over the nails of the three
   central hind digits. The eyes are large and
   the ears large and rounded. The tail is
   rather short and covered with short hairs.

2. The skull is massive and narrow in the facial
   region with a rounded braincase and low
   occipital crests. The paroccipital processes
   are very short.

● 3. The cheek teeth are hypsodont with persistent
   pulps. The upper teeth have single, deep
   enamel folds on each side which almost meet,
   while the lower teeth have a single outer and
   two inner folds.

4. They are rare and little known and are
   restricted to the Andes from southern Peru
   to northern Argentina. There are two species
   in a single genus.

   Example:- Abrocoma.

anterior

upper left molar 1   lower left molar 1

chinchilla rat
Abrocoma

Family Echimyidae - Spiny rats

Hystricomorpha in which:-

1. They are medium-sized, rat-like, ground-
living, burrowing, arboreal or semi-aquatic
rodents with moderately long legs.   The
● hairs tend to be spiny.   The first digit of
the fore-limb is rudimentary and the hind-limb
has five digits.   The eyes and ears are
moderately large.   The tail is of moderate
length or long and may be well-haired or
sparsely haired.

2. The skull tends to be elongated with moderately
prominent occipital crests.   The paroccipital
● processes are relatively long and curved beneath
the bullae.

● 3. The cheek teeth are brachydont and rooted
with an occlusal pattern of transverse ridges
formed by inner and outer folds of enamel
and tending to form islets with wear.

4. They are restricted to tropical South America,
Martinique and Puerto Rico.   There are 43
species in 14 genera.

Examples:-   Echimys, Proechimys, Dactylomys.

spiny rat - Proechimys

anterior

distribution of
the Echimyidae

upper left molar 1

Family Thryonomidae - Cutting grass or cane rats

Hystricomorpha in which:-

1. They are medium-sized to large ground-living and burrowing rodents of heavy build with coarse rather spiny hair. The limbs are relatively short with four digits on both fore-limbs and hind-limbs, the fifth digit being vestigial. The eyes and ears are small. The tail is short and covered with hair.

2. The skull is massive with prominent occipital and sagittal crests and long paroccipital processes.

● 3. The cheek teeth are hypsodont and rooted with an occlusal pattern of transverse ridges formed by inner and outer enamel folds.

4. They are widespread in Africa south of the Sahara. There are 6 species in a single genus.

  Example:- Thryonomys.

posterior

upper right molar 1

cutting grass - Thryonomys

Family Petromyidae - Rock rat

Hystricomorpha in which:-

1. They are relatively small, squirrel-like
   rodents living in rock crevices and capable
   of dorsoventral compression.   The legs are
   relatively short with the first digit,
   vestigial in the fore-limb and very short in
   the hind-limb.   The eyes are of moderate
   size and the ears short.   The tail is of
   moderate length with scattered long hairs
   and a terminal tuft.

● 2. The skull tends to be elongated, broad and
   flat with well-developed occipital and lateral
   crests.   The paroccipital processes are long
   and slender and curved forward around the
   bulla. (108)

● 3. The cheek teeth are hypsodont and rooted with
   a simple occlusal pattern arising from a
   single fold of enamel on each side: The
   internal side of each upper tooth and the
   external side of each lower has two elevations.

4. They are found in Southwest Africa.   There
   is a single species.

   Example:- Petromys

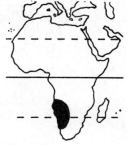

posterior

lower left molar 1

rock rat - Petromys

145

Family Bathyergidae - Mole rats

Hystricomorpha in which:-

● 1. They are small to medium-sized, mole-like
rodents living in systems of shallow tunnels.
The legs are short and strong and pentadactyl.
The eyes and ears are very small and the
tail short.

2. The skull is relatively short and strong with
prominent occipital and sagittal crests.
The paroccipital processes are short.   There
● is a small infraorbital foramen that transmits
little or no masseter muscle.

● 3. The cheek teeth are hypsodont and rooted.
There is a simple occlusal pattern with a
single enamel infold on each side (8-shaped)
in the anterior teeth and none in the posterior
teeth (ring-shaped).   The number of teeth is
reduced according to the genus.

4. They are found in Africa south of the Sahara.
There are 22 species in 5 genera.

Examples:-   Bathyergus, Cryptomys, Georychus,
Heterocephalus.

 posterior

upper left molars 1 & 2

distribution of
the Bathyergidae

blesmol - Georychus

146

Family Ctenodactylidae - Gundis

Hystricomorpha in which:-

1. They are rather small, compact, ground-living
   rodents adapted for rocky, arid conditions.
   The legs are short and the digits reduced to
   four on each foot, the two inner digits of
   ● the fore-limb bearing a comb of bristles.
   The eyes are relatively large and the ears
   of moderate size and rounded.   The tail is
   short and fully haired.

2. The skull is flattened dorsoventrally and
   broad posteriorly with low occipital ridges.
   The paroccipital processes are very short.

● 3. The cheek teeth are hypsodont with persistent
   pulps.   The occlusal pattern is very simple
   with broad relatively shallow infoldings of
   enamel on each side or on one side only resulting
   in an 8-shaped or kidney-shaped grinding surface.

4. They are found in North Africa and the northern
   part of the Sahara.   There are 8 species in
   4 genera.

   Examples:- <u>Ctenodactylus</u>, <u>Pectinator</u>.

posterior

<u>upper right premolar 4</u>
<u>and molar 1</u>

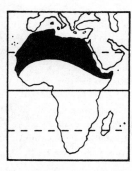

<u>distribution of</u>
<u>the Ctenodactylidae</u>

gundi - <u>Ctenodactylus</u>

147

7   Whales

There are three orders of aquatic mammals, the seals and walruses (Pinnipedia), the dugongs and manatees (Sirenia) and the dolphins and whales (Cetacea). Many other orders, such as the Insectivora, Rodentia and Carnivora, have semi-aquatic members, but all are also well-adapted for life on land. Walking and running in four-footed mammals, with the legs rotated ventrally and thus vertically orientated (see page 17), involve undulation of the body also in the vertical plane. This can be seen, for example, in a running dog, or a cheetah and in the horse when jumping. The phocid seals, which use their limbs mainly for swimming, move on land with caterpillar-like dorso-ventral undulations. This is in contrast with most reptiles, many amphibia and the great majority of fishes where body waves in locomotion are transverse. The tail in swimming forms such as fishes, newts and the extinct ichthyosaurs, therefore, is laterally flattened. In aquatic mammals, since the dorso-ventral body undulation of their land-living ancestors has been retained, the tail flukes (or paddle) are horizontal. This is the case in the Cetacea and Sirenia and even in the Pinnipedia, where the tail is short, the webbed hind limbs, held backward and together, form a horizontal paddle for swimming and move mainly vertically.

The whales and dolphins have the greatest adaptation for life in water of any mammal. They are found in some of the larger rivers and estuaries and particularly on the high seas. They never come on land (and in fact cannot survive if stranded) and exploit the breadth and

depths of the oceans for food.   Deep diving by air-
breathing animals calls for extreme structural and
physiological adaptation.   Prolonged submergence is
necessary for food searching and in the main requires
oxygen to release the energy for muscle contraction and
other cellular processes.   Gaseous air taken down in the
lungs under high pressure passes into solution in the body
fluids and, whereas the oxygen is used, the nitrogen
comes out of solution, on return to the surface, as
bubbles in the blood which cause blockage of capillaries.
The problem is solved in the whales (and also in the
seals) first by myoglobin, a pigment in the muscles,
which serves as a large oxygen store and, secondly, by
reducing both the air content of the blood and its rate
of flow to the tissues.   This is achieved by emptying
the lungs, isolating the blood in sinuses and special
organs and reducing the heart beat to extremely low
levels.   In addition, sensitivity to carbon dioxide is
also reduced to delay the onset of breathing movements.
Anaerobic respiration takes over when oxygen supplies are
exhausted, the oxygen debt being repaid when the animal
surfaces.

In the whales there may be group cooperation.
Whales communicate vocally and use echolocation as a
main sense enabling them to distinguish even the surface
textures of different fishes.   There is now much evidence
from captive and trained dolphins and killer whales that
they have language and powers of reasoning.   They appear
to instruct one another by means of sounds and a case is
known where a dolphin was trained to devise new tricks.
Whale 'songs' are sequences of thousands of sounds which
are repeated.   It seems that they may have oral traditions
and a social life that we do not yet understand.

The whales and Man each in their way represent a
peak of achievement in evolution.   It now appears that
they are twin intelligences on this planet.   It might
have been supposed that there would be no room for two
kinds of intelligent being, except that they occupy
entirely different environments, since inevitably in
competition one must oust the other.   Such competition
did not exist until Man became the super-predator of the
seas.

F

W H A L E S

The commercial hunting of whales dates from the
Middle Ages when Basque seamen harpooned right whales,
Balaena, migrating down the coasts of France and Spain.
By the end of the 17th century shore and shallow water
whaling had developed in northern Europe and the east
coast of North America and rather earlier in Japan. Deep
sea whaling from ocean-going sailing ships was essentially
a 19th century enterprise when whale oil was used for
lamps. But it was the development in recent times of
powered catcher boats using explosive harpoons and
operating from a factory ship that has done most damage
to whale stocks. By the 1940s many of the great-whale
populations had been seriously depleted. This led to the
formation in 1946 of the International Whaling Commission
responsible for the conservation of whales by imposing
hunting·bans on some species, kill quotas on others, size
limits, catch monitoring and the exchange of information.

The range in form among mammals is so great that
mammalogists have boggled at giving a fair and balanced
interpretation of their relationships within the
framework of the Linnaean hierarchy of class, order,
family. Accordingly various superordinal categories
such as 'cohort' have been proposed from time to time by
different authorities. These have served to highlight
useful groupings though not always with general agreement.
For example, Simpson (Bull. Amer. Mus. nat. Hist. 85, 1945)
placed the range of orders commencing with the Insectivora
and passing through the Primates to the edentates in the
Cohort Unguiculata, the rodents and lagomorphs in the
Glires and the Cetacea alone in emphasis of an isolated
position due to extreme specialisation in the Cohort
Mutica. But views change and recently Minkoff
(Zoo. J. Linn. Soc. 58, 1976) included the lagomorphs,
rodents and carnivores in the Unguiculata and placed the
Cetacea in a new Cohort Ungulata with the ungulate Eutheria.

The validity of the Cohort as a taxon is hard to
establish from matching characters of the various orders
as almost none remains (after adaptive radiation) to
point unequivocally to a common origin. In this guide
to living mammals, therefore, the use of superordinal
categories other than class and subclass is avoided.

W H A L E S

Order Cetacea

Eutheria in which:-

1. They are medium-sized to very large exclusively aquatic, streamlined and essentially hairless mammals. They are either carnivores or plankton feeders according to the group.

● 2. The fore-limbs are modified as paddle-shaped flippers (154) without externally visible digits or claws.   Hind-limbs are absent and the pelvis vestigial.   There is a horizontal tail fin.

3. Teeth may be present (Odontoceti) or absent (Mysticeti).

● 4. The skull is highly modified and the bones telescoped. The nasal openings are far back on the dorsal surface (except in Physeteridae).   The parietals do not meet in the midline being separated by the very large supra-occipital and an interparietal.   The maxilla extends back to overlap a considerable part of the frontal.   The cranium is large. (152, 154)

5. There is no postorbital bar.

6. The form of the lachrymal varies according to the group.

7. There is no paroccipital process. (154)

ORDER	SUBORDER
CETACEA	ODONTOCETI
	MYSTICETI

the killer whale - Orcinus

151

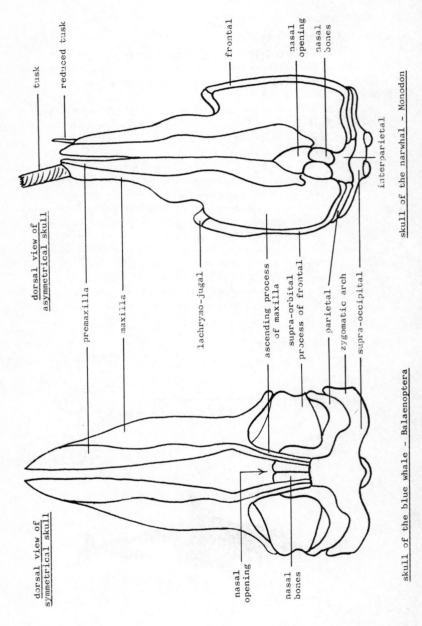

dorsal view of asymmetrical skull

tusk

reduced tusk

frontal

nasal opening

nasal bones

interparietal

skull of the narwhal – Monodon

premaxilla

maxilla

lachrymo-jugal

ascending process of maxilla

supra-orbital process of frontal

parietal

zygomatic arch

supra-occipital

dorsal view of symmetrical skull

nasal opening

nasal bones

skull of the blue whale – Balaenoptera

152

WHALES

Suborder Odontoceti - Toothed Whales

Cetacea in which:-

● 1. Teeth are always present and are either numerous,(154) uniform and conical or are reduced to a single tooth.
● 2. The skull is asymmetrical.(152)
3. The maxilla does not have an orbital plate. Its ascending process is large and broad, does not interlock with the frontal, but spreads out over the supra-orbital process.(152, 154)
● 4. The nasal opening is single.(152)
5. The nasal bones do not form part of the roof of the nasal passage.(152)
6. The lachrymal is either inseparable from the very slender jugal or is very large and forms part of the roof of the orbit.(154)
7. The tympanic is not fused with the periotic.
8. The anterior ribs are two-headed (bicipital).
9. The sternum is composed of three or more parts articulating with three or more ribs.

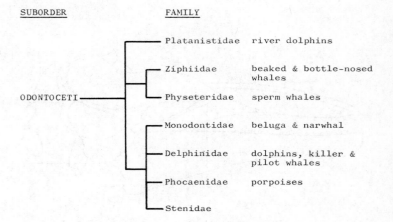

SUBORDER	FAMILY	
	Platanistidae	river dolphins
	Ziphiidae	beaked & bottle-nosed whales
ODONTOCETI	Physeteridae	sperm whales
	Monodontidae	beluga & narwhal
	Delphinidae	dolphins, killer & pilot whales
	Phocaenidae	porpoises
	Stenidae	

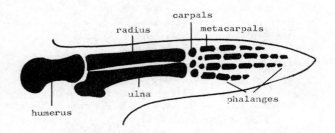

fin of killer whale - Orcinus

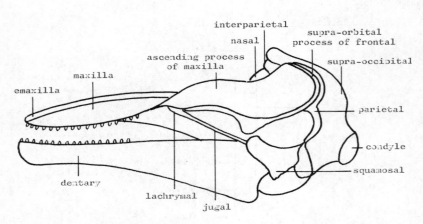

lateral view of skull of the porpoise - Phocaena

W H A L E S

Family Platanistidae - River dolphins

Odontoceti in which:-

- 1. The beak is long and slender, the forehead bulges
     and there is a more or less distinct neck.
     The dorsal fin is low and the flippers short
     and broad. The posterior margin of the
     tail is not deeply notched at the midline.
     They attain a length of 1.5 - 2.9 m.
- 2. The cervical vertebrae are nót fused and are
     long for a cetacean.
- 3. The teeth range from 26/26 (Inia) to 55/55
     (Pontoporia).
  4. They are black (Platanista), brown (Pontoporia)
     or whitish (Inia, Lipotes). Platanista
     occurs in the Indus, Ganges and Brahmaputra
     rivers, is blind and feeds on shrimps and
     bottom-living fishes. Inia is restricted
     to the Amazon and Orinoco rivers of tropical
     South America and feeds exclusively on fish.
     Pontoporia lives in the La Plata estuary and is
     also found along the southeast coast of South
     America taking herring, prawns and squid.
     Lipotes occurs only in Lake Tungt'ing on
     the Yang-Tse-Kiang in China and feeds on
     bottom-living catfish. There are 4 species
     in 4 genera.

     Examples:- Platanista, Pontoporia, Inia,
                 Lipotes.
                            (see page 156)

La Plata dolphin - Pontoporia

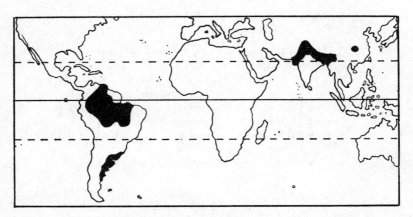

Platanistidae

Monodontidae

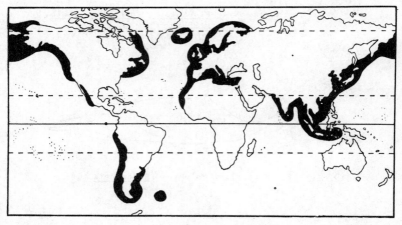

Phocaenidae

W H A L E S

<u>Family Ziphiidae</u> - Beaked and bottle-nosed whales

Odontoceti in which:-

● 1. The snout is narrow and beak-like and in some
(<u>Hyperoodon</u>, <u>Berardius</u>) there is a high bulging
forehead.   The dorsal fin is small and placed
far back and the flippers are also small.
The posterior margin of the tail is not
deeply notched at the midline.   They attain
a length of 4.0 - 12.4 m.

● 2. At least the first two cervical vertebrae
are fused and in <u>Hyperoodon</u> all are fused.

● 3. The teeth are typically reduced to a single
pair or two pairs in the lower jaw (0/1 or
0/2), but in <u>Tasmacetus</u> the dental formula
is about 19/27.

  4. They are typically brown or grey, deep-
diving whales feeding mainly on squid and
some deep sea fishes.   They are found in
all oceans.   There are 18 species in 5 genera.

Examples:-   <u>Ziphius</u>, <u>Berardius</u>, <u>Hyperoodon</u>,
<u>Mesoplodon</u>, <u>Tasmacetus</u>.

(see page 158)

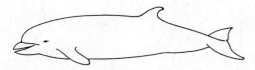

bottle-nosed whale - <u>Hyperoodon</u>

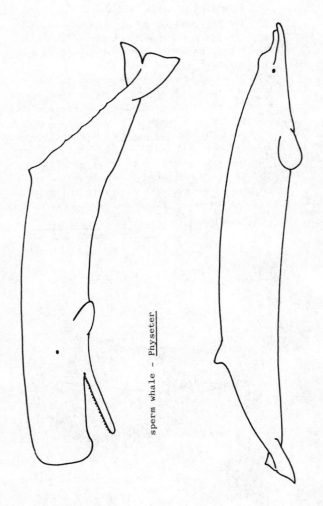

sperm whale - Physeter

Baird's beaked whale - Berardius

W H A L E S

<u>Family Physeteridae</u> - Sperm whales

Odontoceti in which:-

● 1. The head is blunt and more or less rectangular
and very large (<u>Physeter</u>) or rounded and
smaller (<u>Kogia</u>).   The dorsal fin is low and
rounded (<u>Physeter</u>) or slightly hooked (<u>Kogia</u>).
The flippers are broad and relatively short.
The posterior margin of the tail is deeply
notched at the midline (<u>Physeter</u>) or not
(<u>Kogia</u>).   They attain a length of 2.1 -
3.4 m (<u>Kogia</u>) and 8.8 - 18.2 m (<u>Physeter</u>).

● 2. All cervical vertebrae are fused except the
atlas in <u>Physeter</u>.

● 3. The teeth are restricted to the lower jaw,
0/8-16 (<u>Kogia</u>) and 0/25 (<u>Physeter</u>).

  4. They are typically grey, deep-diving whales
feeding mainly on squid, but <u>Physeter</u> also
takes large fish and <u>Kogia</u> takes Crustacea.
They are found in all oceans.   There are
three species in two genera.

Examples:- <u>Physeter</u>, <u>Kogia</u>.

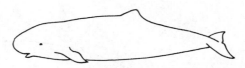

pygmy sperm whale - <u>Kogia</u>

W H A L E S

<u>Family Monodontidae</u> - Beluga and narwhal

Odontoceti in which:-

● 1. They have a short, broad snout (<u>Delphinapterus</u>)
   or none (<u>Monodon</u>) with a high rounded forehead.
   There is no dorsal fin.   The flippers are
   short, broad and rounded.   The posterior
   margin of the tail is more or less deeply
   notched at the midline.   They attain a body
   length of about 6.0 m.

● 2. The cervical vertebrae are not fused.

● 3. The teeth are 9/9 in <u>Delphinapterus</u> and 1/0
   in <u>Monodon</u>.   In the male narwhal (<u>Monodon</u>)
   the single tooth of the left side (occasionally
   the right) develops into a straight tusk up to
   2.7 m long and spirally grooved to the left.

  4. They are grey, white or mottled whales.
   They are not deep divers and feed on a variety
   of benthic and midwater animals including
   squid, Crustacea and fish.   They are
   widespread in Arctic waters and enter the
   larger rivers of Siberia, Canada and Alaska.
   There are two species in two genera.

Examples:- <u>Delphinapterus</u>, <u>Monodon</u>.

(see pages 152, 156)

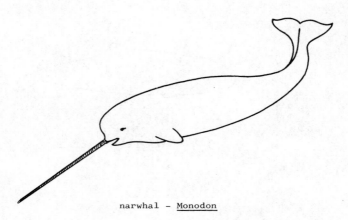

narwhal - <u>Monodon</u>

W H A L E S

<u>Family Delphinidae</u> - Dolphins, killer whales and
pilot whales

Odontoceti in which:-

● 1. They either have a prominent beak
   (Lissodelphinae, Delphininae, except <u>Grampus</u>)
   or none, and usually a more or less bulging
   forehead.   There is a prominent dorsal fin
   (except in Lissodelphinae) usually hooked
   and placed relatively far forward.   The
   flippers may be tapering or paddle-shaped.
   The posterior margin of the tail is more or
   less notched at the midline.   They attain
   a length of 1.5 m in the smaller dolphins
   to 9.4 m in <u>Orcinus</u>.

● 2. The first two cervical vertebrae at least
   are fused, but more (first four in <u>Orcinus</u>)
   or all may be fused (<u>Grampus</u>).

● 3. The teeth are conical and range from 0/2
   (<u>Grampus</u>) to 65/58 (<u>Delphinus</u>).

  4. The colour varies from uniform black or grey
   to patterns of black and white according to
   species.   They do not make excessively deep
   or prolonged dives and feed on a wide variety
   of prey which, in the case of killer whales,
   <u>Orcinus</u>, includes other cetaceans, seals,
   large fishes and diving birds.   They are
   found in all oceans.   There are about 32
   species in 14 genera.

   Examples:- <u>Delphinus</u>, <u>Grampus</u>, <u>Tursiops</u>,
            <u>Cephalorhynchus</u>, <u>Lissodelphis</u>,
            <u>Globicephala</u>, <u>Orcinus</u>.

(see pages 151, 154)

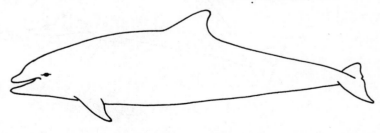

bottle-nosed dolphin - <u>Tursiops</u>

WHALES

Family Phocaenidae - Porpoises

Odontoceti in which:-

● 1. There is no distinct beak, the head being conical or rounded. The dorsal fin is either low and obtuse or absent, and the flippers relatively short. The tail fin is narrow and shallowly notched at the midline. They attain a length of 1.6 - 2.1 m.

● 2. The first three cervical vertebrae are fused (Neophocaena) or six or all seven are fused (Phocaena, Phocaenoides).

● 3. The teeth are characteristically laterally compressed and spade-like with two or three cusps weakly developed on the crown. They range from 15/15 to 30/30.

4. They are either uniformly black or black and white. They are found in estuaries, coastal waters and the open sea throughout the northern hemisphere and also in temperate South American waters. They mainly feed on fish, but take a wide range of prey including squid and Crustacea. There are 7 species in 3 genera.

Examples:- Phocaena, Neophocaena, Phocaenoides.

(see pages 154, 156)

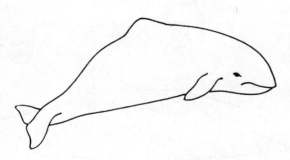

porpoise - Phocaena

Family Stenidae

Odontoceti in which:-

● 1. The snout is long merging more or less
     smoothly into the forehead.   The dorsal fin
     is hooked to a greater or lesser extent and
     the flippers broad at the base.   The tail
     is shallowly notched at the midline.   They
     range from 1.6 - 2.5 m.
● 2. The first two cervical vertebrae are fused.
● 3. The teeth range from about 24/24 (Steno)
     to 32/32 (Sousa and Sotalia).   The teeth
     are rugose in Steno.
   4. They are black, white or speckled.   They
     feed mainly on fish and squid.   Steno is
     found in tropical and warm temperate waters
     in all oceans.   Sotalia occurs in the Amazon
     and Orinoco rivers and the South American
     coast from southern Brazil to Venezuela.
     Sousa occurs in coastal areas of southern
     Asia from China to Zanzibar and West Africa.
     There are 8 species in 3 genera.

   Examples:-   Steno, Sousa, Sotalia.

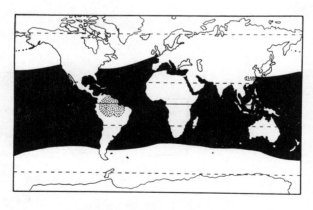

distribution of the Stenidae

W H A L E S

Suborder <u>Mysticeti</u> - Baleen whales

Cetacea in which:-

- 1. Teeth are absent (vestigial in the foetus). Transverse sheets of comb-like baleen (keratin) descend from the roof of the mouth into the buccal cavity and serve to strain plankton. (165)
- 2. The skull is symmetrical. (152)
  3. The maxilla has an orbital plate. Its ascending process is small and narrow, interlocks with the frontal and does not spread out over the supra-orbital process. (152)
- 4. There is a pair of nasal openings.
  5. The nasal bones form the roof of the anterior part of the nasal passage. (152)
  6. The lachrymal is small and distinct from the slender jugal.
  7. The tympanic is fused with the periotic.
  8. The ribs are single headed.
  9. The sternum is a single structure articulating with the first pair of ribs only.

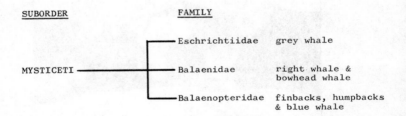

SUBORDER	FAMILY	
	Eschrichtiidae	grey whale
MYSTICETI	Balaenidae	right whale & bowhead whale
	Balaenopteridae	finbacks, humpbacks & blue whale

W H A L E S

Family Eschrichtiidae - Grey whale

Mysticeti in which:-

● 1. The body is slender with a blunt snout and
typically two short grooves on the throat.
There is no dorsal fin, but a low dorsal
hump. The flippers are broad and rather
short. The posterior margin of the tail is
moderately deeply notched at the midline and
there is a crenulate dorsal ridge. They
attain a length of 11.0 - 15.0 m.
● 2. The cervical vertebrae are not fused.
● 3. The baleen plates are short and narrow and
the left and right rows are separated
anteriorly.
4. These whales are grey with white mottling.
They are found in the northern waters of the
Pacific and feed on benthic amphipods.
They migrate in the autumn to Baja California
and Korea where the calves are born in quiet
lagoons and bays in January and February.
They return north in the spring. There is
a single species.

Example:- Eschrichtius.

(see pages 166, 167)

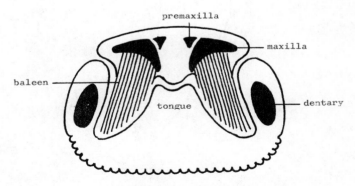

T.S. buccal region of a whalebone whale
showing the baleen plates

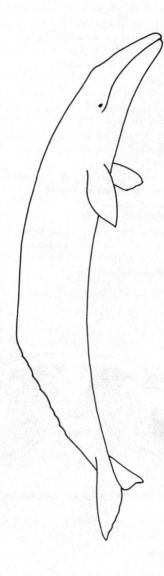

grey whale – <u>Eschrichtius</u>

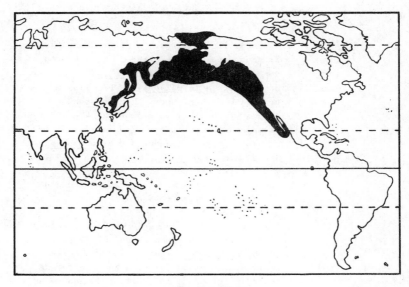

Eschrichtiidae

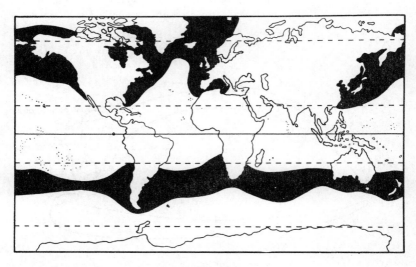

Balaenidae

W H A L E S

<u>Family Balaenidae</u> - Right whale and bowhead whale

Mysticeti in which:-

● 1. The body is robust with a blunt snout and
     without grooves on the throat.   A hooked
     dorsal fin is present in <u>Caperea</u>; there is
     no dorsal fin in <u>Balaena</u>.   The flippers are
     broad and rather short.   The posterior
     margin of the tail is notched at the midline and
     the dorsal ridge is not crenulated.   They
     attain a length of about 6.0 m in <u>Caperea</u>
     and 20.0 m in <u>Balaena</u>.
● 2. The cervical vertebrae are fused.
● 3. The baleen plates are long and narrow and
     the left and right rows are separated
     anteriorly.
  4. These whales are dark grey to black.   They
     feed on plankton in northern waters (bowhead
     whale, <u>Balaena</u> <u>mysticetus</u>) and sub-Antarctic
     seas, Atlantic and northern Pacific (right
     whale, <u>B. glacialis</u>).   Little is known about
     the habits of the pygmy right whale (<u>Caperea</u>)
     which is confined to the southern hemisphere.
     There are 3 species in 2 genera.

Examples:- <u>Balaena</u>, <u>Caperea</u>.

(see pages 167, 169)

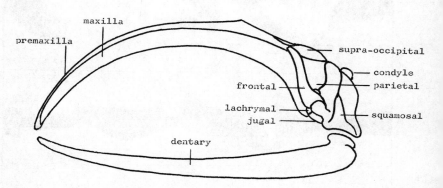

<u>lateral view of skull of the right whale - Balaena</u>

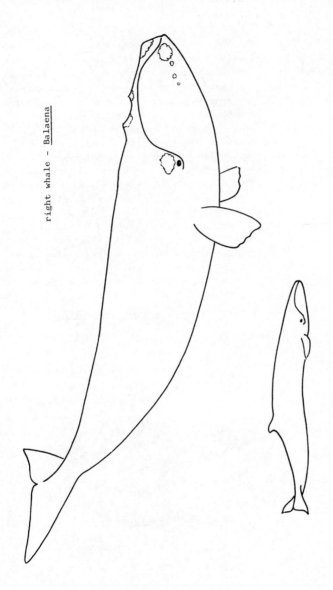

right whale – Balaena

pygmy right whale – Caperea

W H A L E S

Family Balaenopteridae - Finbacks, humpbacks and
blue whale

Mysticeti in which:-

- 1. The body is streamlined with a pointed snout.
  A large number of parallel grooves cover the
  throat and chest.   There is a hooked dorsal
  fin placed rather far back.   The flippers
  are long and narrow.   The tail is slightly
  notched at the midline and irregularly
  scalloped along the posterior edge.   They
  attain a length of about 7.0 - 31.0 m.
- 2. The cervical vertebrae are not fused.
- 3. The baleen plates are short and broad and
  the left and right rows are joined anteriorly.
- 4. These whales are typically grey dorsally with
  a pattern of white on the ventral surface,
  though Megaptera, the humpback, may be
  uniformly black with white flippers.   They
  feed mainly on euphausids (krill) but also
  take small shoaling fishes.   They are found
  in all oceans.   There are 6 species in
  2 genera.

  Examples:-   Balaenoptera, Megaptera.

(see pages 152, 171)

blue whale - Balaenoptera

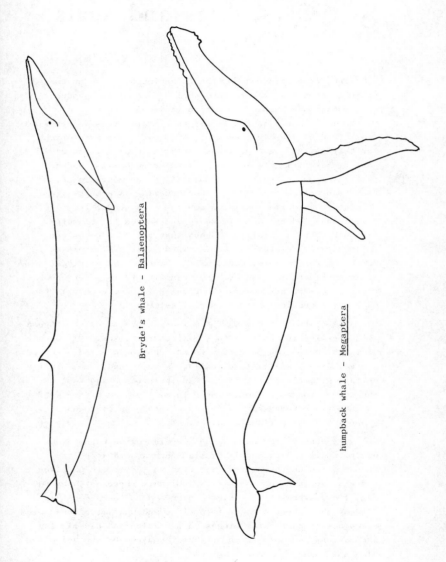

Bryde's whale – <u>Balaenoptera</u>

humpback whale – <u>Megaptera</u>

8   Cats, dogs and seals

The flesh eaters among the Eutheria, with one or two exceptions such as the toothed whales, are in the groups Carnivora and Pinnipedia and here it is rare to find a vegetarian.   This is not to say that meat-eaters do not take plant food or plant-eaters meat.   However, the adaptations for tearing the flesh of larger animals involving powerful jaws with shearing teeth, such as the carnassials, and a particular gut structure and physiology for meat digestion, tend to preclude the use of plant material as a major item of diet.   The carnivores contain the terrestrial forms (Fissipeda) and the aquatic seals and walrus (Pinnipedia). Although there is really no doubt that the pinnipedes are quite closely related to the fissipedes, it has become a matter of convenience, particularly here in the preparation of sets of matching characters, to split off this very compact group as a separate order, the Pinnipedia, and to retain the name Carnivora for the terrestrial animals.

There are two groups of families in the Carnivora, those with dog-like and those with cat-like affinities.   In both of these groups the primitive forms (Procyonidae and Viverridae) are relatively small, long-bodied, short-legged, long-skulled animals with clear resemblances in their teeth and feet to the insectivores.   The advanced forms typified by the wolf and the tiger represent two kinds of predator, the hunting group and the individual stalker.

Whenever in the past groups developed the carnivorous habit, as in the extinct marsupial Borhyaenoidea and the Creodonta which themselves gave rise to the modern Carnivora, types arose in each case similar to those represented by the dog, lion, sabre-toothed tiger, bear and so forth.   It is evident that these extinct tiger-like and bear-like forms were occupants of particular niches in predation and had similar adaptations and body forms to those required by the animals that fill these niches today.

Order Carnivora

Eutheria in which:-

1. They are small to large terrestrial, arboreal or amphibious mammals with plantigrade to digitigrade gait. They typically feed on other animals. (45)
2. The limbs have four or five digits with typically curved, sharp claws.   The metacarpals and metatarsals are never elongated. (174)
● 3. There are three incisors in each half of the upper and lower jaws (except in Enhydra), the third being the largest.   The canine is well developed and the premolars and molars tend to be reduced in number and have crushing and cutting surfaces. (174)
● 4. The skull is robust with a rounded brain case often with prominent crests and a strong outwardly bowed zygomatic arch. (174)
5. There is no postorbital bar.
6. The lachrymal does not extend onto the face and is vestigial in the Ursidae. (174)
7. There is a well-developed paroccipital process. (174)

ORDER   FAMILY

```
                                      ┌─ Procyonidae
                                      │                    ┐
                              ┌───────┤─ Ursidae           │ Canoid
                              │       │                    │ Families
                              │       ├─ Mustelidae        │
                              │       └─ Canidae           ┘
CARNIVORA ────────────────────┤
                              │       ┌─ Viverridae        ┐
                              │       │                    │ Feloid
                              └───────┤─ Hyaenidae         │ Families
                                      └─ Felidae           ┘
```

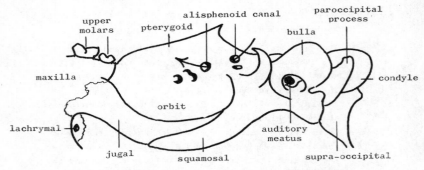

ventro-lateral view of the orbit of a dog skull

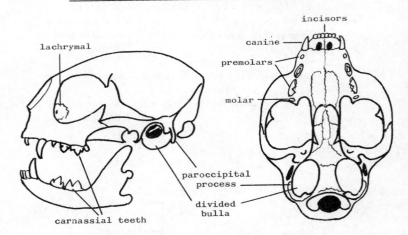

lateral and palatal views of a cat skull

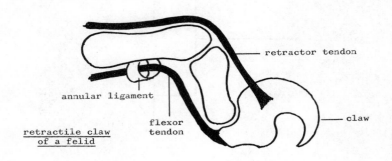

retractile claw
of a felid

174

Family Procyonidae - Racoons

Carnivora in which:-

1. They are relatively short-legged, tree-climbing animals with a plantigrade or semi-plantigrade gait.(45)
● 2. The claws are not retractile or are partially retractile.(174)
● 3. The canines are long and rectangular in section. The premolars are pointed and the molars tend to have rounded cusps (bunodont). Carnassial teeth for cutting (last upper premolar and first lower molar) are weakly developed. The dental formula is usually I 3/3, C 1/1, Pm 4/4, M 2/2.
● 4. The skull is moderately long.
5. There is no alisphenoid canal except in the panda, Ailurus.(174)
6. There is a prominent paroccipital process.
7. The auditory bulla is well rounded.
8. They are omnivorous except for the giant panda, Ailuropoda, which is vegetarian.
9. They are found in most of North and South America and in a small area of southeast Asia. There are 8 genera and 19 species.

Examples:- Ailurus, Nasua, Procyon, Ailuropoda.

N.B. Some authorities place the giant panda, Ailuropoda, with the Ursidae or in a separate family.

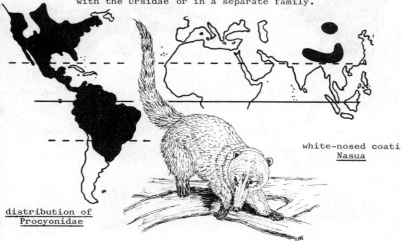

white-nosed coati
Nasua

distribution of
Procyonidae

Family Ursidae - Bears

Carnivora in which:-

● 1. They are relatively slow-moving animals with large powerful legs and a plantigrade gait.(45)

● 2. The claws are not retractile.

3. The canines are long and the cheek teeth have rounded cusps mainly for crushing. The carnassial teeth (last upper premolar and first lower molar) are weakly developed. The dental formula is typically I 3/3, C 1/1, Pm 4/4, M 2/3.

4. The skull is long.

● 5. There is an alisphenoid canal.(174)

6. The paroccipital process is large and broad.

● 7. The auditory bulla is flattened.

8. They are omnivorous except for the polar bear.

9. They are found throughout the northern hemisphere and are absent from the greater part of South America and from Africa and Australasia. There are 5 genera and 7 species.

Examples:- Ursus, Melursus.

distribution
of Ursidae

the black bear
Ursus

Family Mustelidae - Badgers, Otters and Weasels

Carnivora in which:-

● 1. They are short-legged, long-bodied animals and include burrowing, tree climbing and amphibious forms.   The gait is plantigrade to digitigrade.(45)
● 2. The claws are partially retractile.
3. The canines are long and sharp.   The cheek teeth may be reduced in number and tend to have sharp cusps.   Carnassial teeth for cutting (last upper premolar and first lower molar) are usually well developed.   The dental formula is usually I 3/3, C 1/1, Pm 3/3, M 1/2.
4. The facial part of the skull is shortened.
5. There is no alisphenoid canal.(174)
● 6. There is a prominent paroccipital process. A postglenoid process locks the mandible in place.
7. The auditory bulla is inflated, but tends to be more or less flattened.
8. They usually eat flesh, but some also eat insects, vegetable material and honey.
9. They are world-wide in distribution except for Madagascar and Australasia.   There are 25 genera and 70 species.

Examples:- <u>Mustela</u>, <u>Mellivora</u>, <u>Meles</u>, <u>Mephitis</u>, <u>Lutra</u>, <u>Enhydra</u>.

weasel - <u>Mustela</u>

177

<u>Family Canidae</u> - Dogs, Wolves, Foxes and Jackals

Carnivora in which:-

1. They are long-legged running animals with
   a digitigrade gait. (45)

● 2. The claws are not retractile and are blunt
   and straight.

3. The canines are long and the cheek teeth have
   rounded cusps for crushing except for the
   carnassials (last upper premolar and first
   lower molar) which are well developed for
   cutting.   The dental formula is usually I 3/3,
   C 1/1, Pm 4/4, M 2/3.

4. The facial part of the skull is long.

● 5. There is an alisphenoid canal. (174)

6. The paroccipital process is long. (174)

7. The auditory bulla is rounded. (174)

8. They are more or less carnivorous, but some
   such as the foxes and jackals also eat vegetation.

9. They are world-wide in distribution except for
   Madagascar and most oceanic islands.   There are
   15 genera and 41 species.

Examples:- <u>Canis</u>, <u>Vulpes</u>, <u>Lycaon</u>, <u>Otocyon</u>.

wolf- <u>Canis</u>

Family Viverridae - Civets, Genets and Mongooses

Carnivora in which:-

● 1. They are short-legged, long-bodied, terrestrial
and arboreal animals with a semiplantigrade to
digitigrade gait.(45)

● 2. The claws are partially retractile.

3. The canines are relatively small and the cheek
teeth have sharp cusps.   The carnassials (last
upper premolar and first lower molar) are not
well developed.   The dental formula is usually
I 3/3, C 1/1, Pm 3-4/3-4, M 2/2.

4. The facial part of the skull is relatively long.

● 5. There is usually an alisphenoid canal.(174)

6. The paroccipital process is in close contact
with the auditory bulla.

● 7. The auditory bulla is inflated with an external
constriction and an internal partition dividing
the cavity.

8. They mostly eat small animals including fish,
crustacea and insects, but some also eat nuts,
fruit and bulbs.

9. They are found in the Old World tropics and
subtropics, but are absent from Australasia.
There are 36 genera and 75 species.

Examples:-   <u>Civettictis</u>, <u>Genetta</u>, <u>Fossa</u>, <u>Viverra</u>,
<u>Paradoxurus</u>, <u>Herpestes</u>, <u>Prionodon</u>.

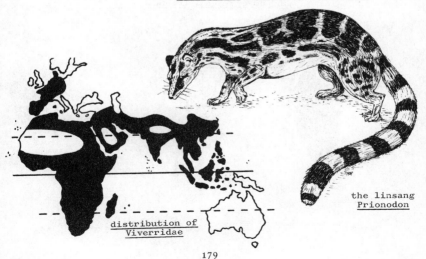

the linsang
<u>Prionodon</u>

<u>distribution of
Viverridae</u>

179

Family Hyaenidae - Hyaenas and Aard-wolf

Carnivora in which:-

- 1. They are terrestrial animals with the hind-limbs shorter than the fore-limbs.   The gait is digitigrade. (45)
- 2. The claws are not retractile and are stout and relatively straight.
  3. The canines are long.   In the hyaenas the cheek teeth are massive with long rounded cusps and well-developed carnassials (last upper premolar and first lower molar).   The dental formula is I 3/3, C 1/1, Pm 4/3, M 1/1.   In Proteles, the aard-wolf, the cheek teeth are reduced and feeble (Pm 3/1-2, M 1/1-2).
  4 The skull is long.
  5. There is no alisphenoid canal.(174)
- 6. The paroccipital process is in close contact with the auditory bulla.
- 7. The auditory bulla is inflated and not divided internally with a partition in hyaenas, but divided in Proteles.
  8. They are mostly carrion feeders, but Proteles also eats insects.
  9. They are found in Africa, southwest Asia and India.   There are 3 genera and 4 species.

Examples:- Hyaena, Crocuta, Proteles.

distribution of
Hyaenidae

brown hyaena - Hyaena

Family Felidae - Cats

Carnivora in which:-

1. They are relatively long-legged, terrestrial
   and arboreal animals with a digitigrade gait.(45)
● 2. The claws are fully retractile.(174)
● 3. The incisors are in a straight transverse line.
   The canines are long and the cheek teeth sharp-
   cusped, powerful and reduced in number.   The
   carnassials (last upper premolar and first lower
   molar) are very well developed for cutting.
   The dental formula is usually I 3/3, C 1/1,
   Pm 3/2, M 1/1. (174)
4. The skull is short and rounded. (174)
5. There is no alisphenoid canal.(174)
● 6. The paroccipital process is flattened against
   the auditory bulla. (174)
● 7. The auditory bulla is highly inflated and
   divided internally with a partition. (174)
8. They typically feed on higher vertebrates though
   some take fish, invertebrates and fruit.
9. They are world-wide in distribution except for
   Madagascar and Australasia.   There are 4 genera
   and 37 species.

Examples:- Felis, Acinonyx, Panthera.

tiger - Panthera tigris

G

Order Pinnipedia

    Eutheria in which:-

1. They are medium-sized to large aquatic mammals frequenting shores and ice flows and returning to land to breed.   They feed on fish, squid and shell-fish and some southern hemisphere species take penguins.

● 2. The digits are fully webbed.   The fore-limbs act as flippers while the hind limbs, backwardly directed in swimming, act as a propulsive 'tail'.   Claws are present but tend to disappear. (183, 184, 186)

● 3. There are fewer than three incisors in each half of the lower jaw.   The canines and cheek teeth are all more or less conical.   The cheek teeth may have accessory cusps before and behind, but are never broad and tuberculated. (183)

4. The cranial part of the skull is large compared (183) with the facial region.   The zygoma is nearly straight.

5. There is no postorbital bar.

6. The lachrymal is absent.

7. The paroccipital process is rudimentary or absent.

ORDER

FAMILY

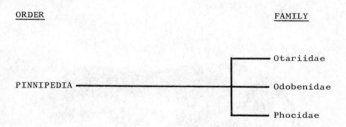

PINNIPEDIA ─────────────────────────────── Otariidae

                                               Odobenidae

                                               Phocidae

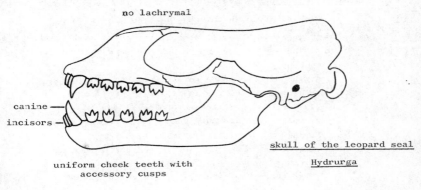

no postorbital bar

no lachrymal

canine

incisors

skull of the leopard seal

Hydrurga

uniform cheek teeth with
accessory cusps

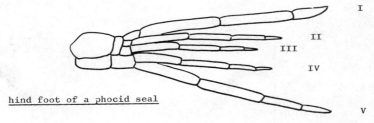

I

II

III

IV

hind foot of a phocid seal

V

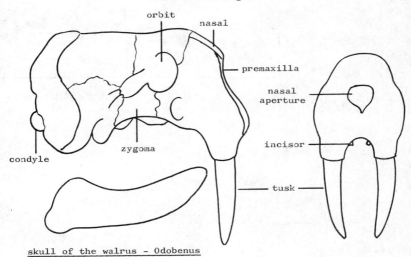

short facial region

orbit

nasal

premaxilla

nasal
aperture

incisor

tusk

zygoma

condyle

skull of the walrus - Odobenus

183

Family Otariidae - Sea Lions

Pinnipedia in which:-

● 1. There is a small external ear and a small tail.
● 2. The hind-limbs are brought forward beneath the body on land.
3. The upper canines do not form tusks.   The dental formula is I 3/2, C 1/1, Pm 4/4, M 1-3/1.
4. There is an alisphenoid canal. (174)
5. The testes are suspended in a scrotum.
6. They feed chiefly on small fish.
7. They are found on Pacific and south Atlantic coasts and on many oceanic islands of the southern hemisphere.   There are 6 genera and 12 species.

Examples:- <u>Otaria</u>, <u>Arctocephalus</u>.

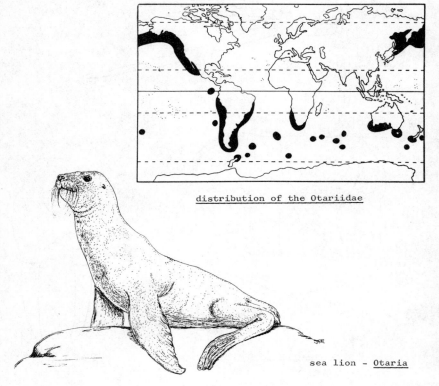

distribution of the Otariidae

sea lion - <u>Otaria</u>

Family Odobenidae - Walrus

Pinnipedia in which:-

- 1. There is no external ear and no free tail.
- 2. The hind-limbs are brought forward beneath the body on land.
  3. The upper canines form very large tusks in both sexes.   The cheek teeth are adapted for crushing. The dental formula is I 1-2/0, C 1/1, Pm 3-4/3-4, M 0/0. (183)
  4. There is an alisphenoid canal. (174)
  5. The testes remain in the abdomen.
  6. They feed on bottom-living molluscs, crustacea and starfish.
  7. They live in the shallow seas and on ice flows at the margin of the Arctic Ocean.   There is 1 genus and 1 species.

Example:- Odobenus.

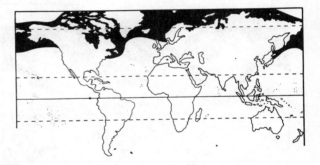

distribution of the Odobenidae

walrus - Odobenus

<u>Family Phocidae</u> - Seals

Pinnipedia in which:-

- 1. There is no external ear but a small tail is present.
- 2. The hind-limbs extend backward on land.
- 3. The upper canines do not form tusks.   The dental formula is I 2-3/1-2, C 1/1, Pm 4/4, M 0-2/0-2. (183)
- 4. There is no alisphenoid canal. (174)
- 5. The testes remain in the abdomen.
- 6. Most species feed on fish, squid and shellfish, but some feed on macroplankton and another on penguins.
- 7. They are found on coasts north of 30° N. latitude and south of 50° S. latitude and also in the Caspian Sea and Lake Baikal.   There are 13 genera and 18 species.

Examples:- <u>Phoca</u>, <u>Monachus</u>, <u>Cystophora</u>, <u>Hydrurga.</u>

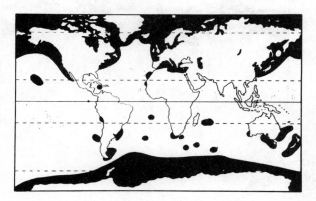

<u>distribution
of the Phocidae</u>

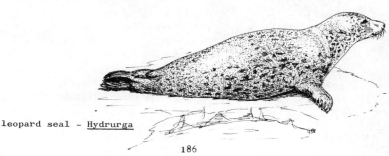

leopard seal - <u>Hydrurga</u>

9 The primitive ungulates

A number of archaic, plant-eating ungulates appeared in the Palaeocene, flourished in both the Old World and the New and then largely became extinct by the end of the Pliocene. However, a few of their relations still exist one of which is the aardvark or Cape anteater. This animal represents a very ancient independent offshoot from the protungulate stock and, apart from its specialised mode of feeding, retains unchanged many of the features present in the early eutherians.

Arising also from early ungulate stock in the Palaeocene are three distantly related groups, the elephants, the hyraxes and the sea cows, each of which traces a separate line back at least to the Eocene. The elephants are a relict of a dying order once numerous on all continents except Australia. They include the largest terrestrial animals that have existed and contrast with the small rabbit-like hyraxes which are perhaps the elephant's closest living relatives. Hyraxes have always been limited to Africa and the Mediterranean region and have remained relatively unchanged since the Oligocene and probably earlier.

The last of these groups, the Sirenia or sea cows is the second mammal order to become exclusively aquatic and to have lost the ability to venture back on to land. Sea cows are known as fossils from the Eocene and have resemblances to the land ungulates of the time. They feed on riverine and coastal vegetation and have never been a large order. The giant of the group, Stella's sea cow of northern waters, was extensively slaughtered and is believed to have become extinct in the 18th Century, but recent unidentified sitings off the coast of Siberia give cause to hope that this may not be so.

Order <u>Tubulidentata</u> , <u>Family Orycteropodidae</u> - the Aardvark

Eutheria in which:-

● 1. They are medium-sized terrestrial and burrowing mammals with digitigrade gait. They feed with a long, protrusible worm-like tongue almost exclusively on termites and ants. (45)

2. The fore-limb has four digits and the hind limb five with shovel-shaped, nail-like hooves for digging.

● 3. The teeth are peg-like and without enamel. Each is composed of numerous hexagonal prisms of dentine surrounding tubular pulp cavities. They grow continuously. The dental formula is I 0/0, C 0/0, Pm 2/2, M 3/3.

● 4. The skull is elongated with a large olfactory region. The palate has a thickened posterior margin and is usually fenestrated. It does not include the pterygoids.

5. There is no postorbital bar.

6. The lachrymal bone is large and extends on to the face.

7. There is no paroccipital process.

8. They are found in Africa south of the Sahara. There is 1 genus and 1 species.

Example:- <u>Orycteropus</u>.

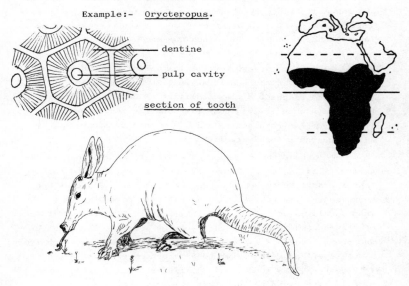

dentine

pulp cavity

<u>section of tooth</u>

aardvark - <u>Orycteropus</u>

THE PRIMITIVE UNGULATES

<u>Order Proboscidea</u>, <u>Family Elephantidae</u> - Elephants

Eutheria in which:-

● 1. They are very large terrestrial mammals with
   digitigrade gait.   They feed on vegetation gathered
   with a long muscular trunk which is an extension of
   the nose with the nostrils at the end. (45, 190)

2. The limbs are pillar-like, each with five toes
   encased in a common integument and ending in
   short hooves. (190)

● 3. The third upper incisor forms a continuously growing
   tusk.   No other incisors or the canines are present.
   There are three milk premolars and three molars, all
   alike with many transverse enamel folds, but only one
   tooth in each half jaw is in use at a time.   It
   moves forward and is shed being replaced from behind
   by the next.   The dental formula is I 1/0, C 0/0,
   Pm 3/3, M 3/3. (190)

4. The skull is short and high with numerous air
   cavities in the bone.   The nasal aperture is high
   on the face. (190)

5. There is no postorbital bar.

6. The lachrymal lies within the orbit.

7. There is no paroccipital process.

8. They occur in Africa south of the Sahara and in the
   Oriental Region.   There are 2 genera and 2 species.

Examples:-   <u>Elephas</u>, <u>Loxodonta</u>.

<u>distribution of the
Proboscidea</u>

African elephant - <u>Loxodonta</u>

189

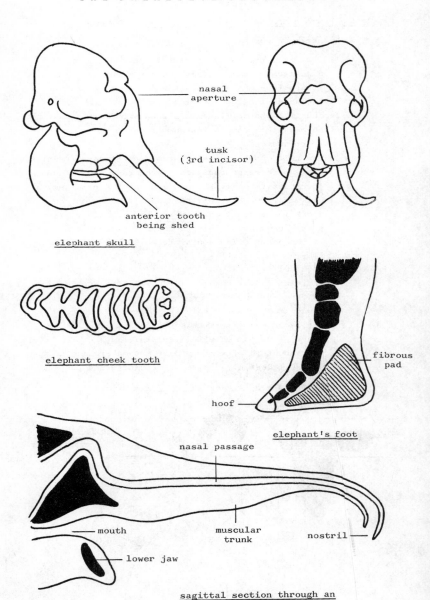

nasal
aperture

tusk
(3rd incisor)

anterior tooth
being shed

elephant skull

elephant cheek tooth

fibrous
pad

hoof

elephant's foot

nasal passage

mouth

muscular
trunk

nostril

lower jaw

sagittal section through an
elephant's trunk

190

Order Hyracoidea, Family Procaviidae - Hyraxes

Eutheria in which:-

- 1. They are small terrestrial or arboreal mammals with a plantigrade gait. They feed on vegetation. They have short ears, no tail and a gland in the centre of the back. (45)

- 2. The fore-limb has four digits and the hind-limb three with short flat hooves, except for the second of the hind-limb which has a long, curved claw.

- 3. There is a single, pointed upper incisor, triangular in section, with enamel on the anterior surface and a persistent pulp and two lower incisors. There is no permanent canine and a wide diastema is present. The cheek teeth form a molariform row with grinding surfaces. The dental formula is I 1/2, C 0/0, Pm 4/4, M 3/3. (192)

- 4. The skull is robust and flattened dorsally. The nasals expand posteriorly and there is a well-developed interparietal. The palate has a thickened posterior margin. (192)

- 5. There is a postorbital bar (sometimes incomplete) formed from the jugal and the parietal. The jugal forms part of the glenoid fossa. (192)

  6. The lachrymal bone is small and is elongated into an antorbital process. (192)

  7. There is a large paroccipital process. (192)

  8. They are found in Africa, the eastern Mediterranean region and Arabia. There are 3 genera and 11 species.

  Examples:- Dendrohyrax, Procavia.

distribution of
the Hyracoidea

the tree hyrax
Dendrohyrax

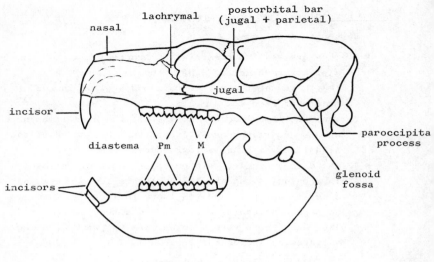

skull of Dendrohyrax

molar of Hyrax

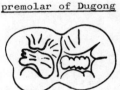

premolar of Dugong

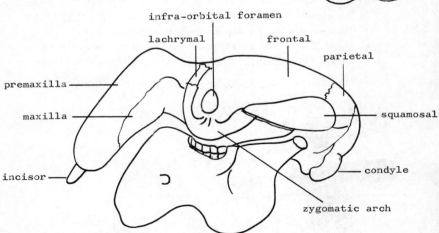

skull of Dugong

Order Sirenia

Eutheria in which:-

● 1. They are medium-sized to large exclusively aquatic mammals with massive, broadly streamlined, thick-skinned, nearly hairless bodies. They feed on aquatic vegetation in coastal waters and rivers in the tropics. (194, 195)

● 2. The fore-limbs are paddle-like. There are no hind-limbs and the pelvis is vestigial. There is a horizontal tail. (194, 195)

3. The teeth differ in the two families.

4. The skull is dense and heavy with roughened bones. The external nares are large and at the level of the orbits.

5. The postorbital bar is incomplete.

6. The lachrymal is small.

7. There is no paroccipital process.

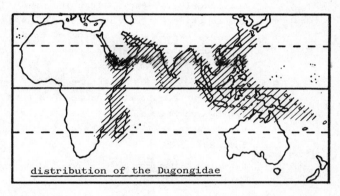

distribution of the Dugongidae

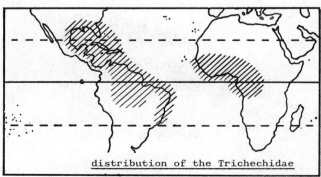

distribution of the Trichechidae

Family Dugongidae - Dugong

Sirenia in which:-

1. They are grey, brown, black or greenish on the back and whitish on the belly. The upper lip is moderately cleft and the nostrils more or less dorsal in position. The tail is notched on the hind margin. They attain a length of 5.8 m.

2. In the skull the premaxillae are much enlarged and deflected downward. The nasal bones are absent, the frontal region broad and the olfactory chamber short. (192)

3. In the lower jaw the mandibular symphysis is deflected downward at an angle of almost 90°, the coronary process is vertical and the condyle is nearly circular. (192)

4. The functional teeth are a single upper incisor, tusk-like in the male, no canines and four to six columnar, cement-covered teeth (I 1/0, C 0/0, Pm 0/0, M 2-3/2-3). (192)

5. There are 59 or 60 vertebrae, with 7 cervicals.

6. They occur in the coastal regions from East Africa and Madagascar to the Philippines, northern Australia and the western Pacific islands. There is a single species.

Example:- Dugong.

the dugong - Dugong

<u>Family Trichechidae</u> - Manatees

Sirenia in which:-

1. They are uniformly dark grey or black.
   The upper lip is deeply cleft, with each
   half capable of independent movement.   The
   nostrils are at the apex of the muzzle.
   The tail is spatulate (without a notch).
   They attain a length of 4.5 m.

● 2. In the skull the premaxillae are relatively
   small and slightly deflected downward.   The
   nasal bones are small and thick, the frontal
   region elongated and the olfactory chamber
   long.

● 3. In the lower jaw the mandibular symphysis
   is deflected downward at an angle of about
   45°, the coronary process slants forward and
   the condyle is elongated transversely.

● 4. There are numerous, low-crowned enamelled
   cheek teeth, without cement and of indefinite
   number which fall out anteriorly and are
   replaced from behind.   There are no incisors
   or canines.

● 5. There are 48 - 50 vertebrae, with 6 cervicals.

6. They occur in the coastal regions and major
   rivers from North Carolina south through
   the Gulf of Mexico and the West Indies to
   latitude 20° S on the South American coast
   and also in West Africa.   There are 3 species
   in a single genus.

Example:- <u>Trichechus</u>.

North American manatee - <u>Trichechus</u>

10   Perissodactyles

The Perissodactyla are the odd-toed ungulates, that is the number of toes is reduced to three or one (rarely four and then the outer one is weak), due to the fact that the weight-bearing axis of the foot passes down the central digit. They were numerous throughout the Tertiary and included the huge horned titanotheres and the clawed chalicotheres which are now extinct. Today the group is represented by three types, the tapir, the rhinoceros and the horse, all reduced to relatively low numbers, at least in the wild, and showing the discontinuous distribution which is characteristic of groups that were once widespread but are now eliminated from much of their former range.

There is little doubt that the tapir and the rhinoceros are more closely related than is either to the horse. This is recognised by dividing the group into two suborders, the Ceratomorpha for the tapir and rhinoceros and the Hippomorpha for the horse, asses and zebras.

The ancestral perissodactyle condition is best shown by the tapir which has changed very little since Oligocene or even Eocene times. It is adapted for living on soft forested ground and eating lush vegetation. The horses and to a lesser extent the rhinoceroses have adapted for faster movement over firm ground. This has involved an increase in body size and length of leg and also the development of a tooth row capable of grinding the hard silicious vegetation of the plains.

The final decline of the order probably dates from the last 20,000 years and is attributable to Man. Cave paintings in Europe made at that time show wild horse and rhinoceros as objects of the chase and there is little doubt that this is the cause of their greatly reduced distribution.

Order Perissodactyla

Eutheria in which:-

1. They are medium-sized to large, terrestrial mammals with an unguligrade gait. They feed on vegetation. (45)

● 2. The axis of the foot passes down the third digit (mesaxonic) which is larger than the others. The digits end in hooves.

3. Some incisors and the canines may be reduced or absent and there is usually a diastema. The cheek teeth form a molariform row with a crushing or grinding surface. (198)

● 4. The skull is elongated through enlargement of the facial bones and the nasal bones project freely for a part or the whole of their length. Horns with a bony core are never present. (198)

5. The postorbital bar when present (Equidae only) contains a process from the squamosal. (198)

6. The lachrymal extends onto the face. (198)

7. There is a well-developed paroccipital process. (198)

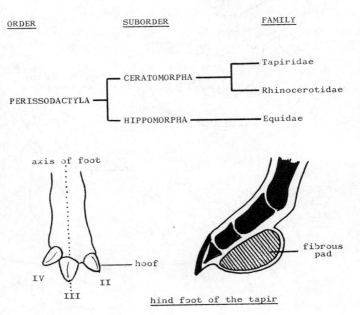

hind foot of the tapir

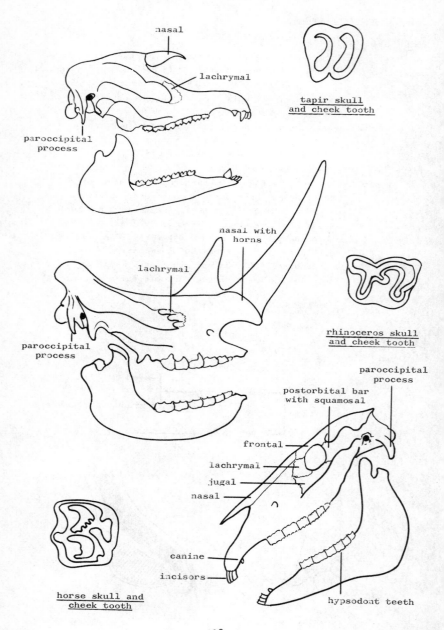

nasal

lachrymal

paroccipital
process

tapir skull
and cheek tooth

nasal with
horns

lachrymal

paroccipital
process

rhinoceros skull
and cheek tooth

paroccipital
process

postorbital bar
with squamosal

frontal

lachrymal

jugal

nasal

canine

incisors

horse skull and
cheek tooth

hypsodont teeth

PERISSODACTYLES

Suborder Ceratomorpha

Perissodactyla in which:-

1. The head is held more or less horizontal. (200)
● 2. The limbs are short and stout with three or four digits on the forefoot and three on the hind.
● 3. The cheek teeth are relatively low crowned with a simple crescentic or a moderately complex pattern.
● 4. The nasal bones are stout and project.

Family Tapiridae - Tapirs

Ceratomorpha in which:-

● 1. The nose is extended into a short, mobile trunk. There are no horns. The body is covered with short hair. The skin is not excessively thick.
2. The dental formula is I 3/3, C 1/1, Pm 4/4, M 3/3.
● 3. There is a short tail without a tuft of hair.
4. They feed on forest vegetation and fruit.
5. They are found in tropical South America and in Burma, Thailand, the Malay Peninsula and Sumatra. There is 1 genus and 4 species.

Example:- <u>Tapirus</u>.

Family Rhinocerotidae - Rhinoceroses

Ceratomorpha in which:-

● 1. The nose does not form a trunk, but bears one or two median horns. The body is sparsely haired and the very thick skin falls into plate-like folds.
2. The dental formula is I 0-2/0-1, C 0/0-1, Pm 3-4/3-4, M 3/3.
● 3. There is a short tail ending in a tuft of bristles.
4. They feed on a variety of vegetation.
5. They are found in Africa and southeast Asia. There are 4 genera and 5 species.

Examples:- <u>Rhinoceros</u>, <u>Diceros</u>.

(see page 200)

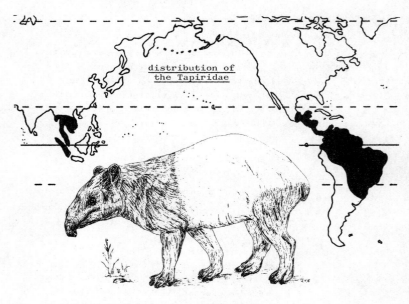

distribution of
the Tapiridae

Malayan tapir - *Tapirus*

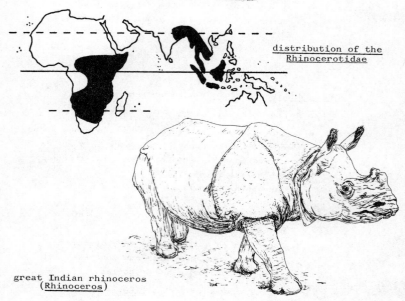

distribution of the
Rhinocerotidae

great Indian rhinoceros
(Rhinoceros)

PERISSODACTYLES

Suborder Hippomorpha, Family Equidae - Horses

Perissodactyla in which:-

1. The head is directed downward.
● 2. The limbs are long and slender with a single hoof borne on the third digit.
● 3. The cheek teeth are very high crowned (hypsodont) and have complex grinding surfaces of exposed enamel, dentine and cement.   The dental formula is I 3/3, C 0-1/0-1, Pm 3-4/3, M 3/3. (198)
● 4. The nasal bones are long and narrow and are free at the anterior end. (198)
5. The body is covered with short hair.   There is a bushy mane and the short tail bears a long whisk of hair.   They feed chiefly on grasses.   Wild species are found in Africa, Arabia and west and central Asia.   There is 1 genus and 7 species.

Example:- Equus.

distribution of the Equidae

African wild ass - Equus

11   Artiodactyles

The artiodactyles are the latest of the ungulates and
currently by far the most successful. They are even-toed with
either four or two digits, the weight-bearing axis passing
between digits three and four with digit one rarely present
even in early forms. Although their expansion has been late
their origin is early and dates from the Eocene. They are
quite distinct from the Perissodactyla and any similarities
between the two groups are due to convergence.

The pigs, chevrotains and llamas represent Eocene types
and have changed very little since that time. They have a
three-chambered stomach but whereas the chevrotain and the
llama ruminate the pig does not. In rumination, food material
after fermentation in the rumen of the stomach is returned to
the mouth as the cud and is chewed a second time. It then
passes back to the reticulum and abomasum of the stomach for
further cellulose digestion. In the higher ruminants, the
deer, giraffes, prongbuck and bovids, a fourth chamber, the
omasum or psalterium is developed between the reticulum and
the abomasum leading to even greater efficiency in the use
of cellulose as a food substrate. This digestive mechanism
making large energy sources available to the animal appears to
have been the key to the success of higher artiodactyles.

But not all ruminants are equally successful. Three
groups, the camels and llamas, the giraffes and the prongbucks
are remnants of a once far greater assemblage. The camels
and the prongbuck originated and diversified in North America
and the giraffe in Africa. On the other hand the deer and
the bovids are at the height of their radiation. The bovids
in particular are undergoing intricate diversification within
three main groups, the sheep and goats, the antelopes and the
cattle and for this reason tend to defy rational classification.

ARTIODACTYLES

Order Artiodactyla

Eutheria in which:-

1. They are small to large terrestrial or amphibious
   mammals with an unguligrade gait, except for the
   Tylopoda which are digitigrade.   They feed on
   vegetation. (45)
● 2. The axis of the foot passes between the third and
   fourth digits (paraxonic).   The first digit is
   absent and the second and fifth usually reduced or
   lost.   The digits end in hooves, those of the third
   and fourth being equal in size and usually flattened
   on the inner and ventral surfaces.
3. Some incisors and the canines may be reduced or absent
   or form tusks.   There is a wide diastema in most
   forms.   The premolars, except occasionally for the
   fourth, are simpler than the molars. (205, 210)
4. The skull is elongated through enlargement of the
   facial bones.   The nasals may or may not project, the
   frontals are usually large and the parietals reduced.
● When horns are present they have a bony core. (205)
5. There may or may not be a postorbital bar formed from
   the frontal and jugal only. (205, 210)
6. The lachrymal is usually large and extends on to the
   face (except in Tayassuidae). (205, 210)
7. The paroccipital process may be large or small. (205, 210)

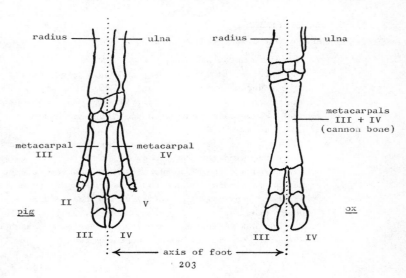

A R T I O D A C T Y L E S

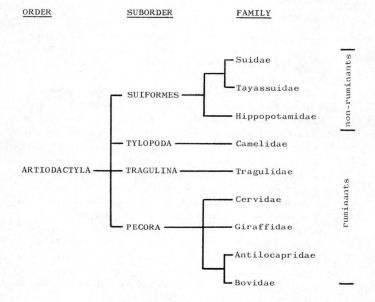

Suborder Suiformes

Artiodactyla in which:-

1. There are no horns.

● 2. The upper canines form tusks and the cheek teeth have rounded cusps (bunodont).

3. The postorbital bar is usually incomplete or absent.

4. There is no vacuity between the lachrymal and nasal bones.

● 5. The feet have four digits except in the Tayassuidae where the hind-feet have three. (203)

6. The third and fourth metapodials are separate except in the Tayassuidae where the metatarsals are fused at their proximal end. They do not form a cannon bone. (203)

7. They do not ruminate. The stomach has three chambers except in the Suidae where there are two.

204

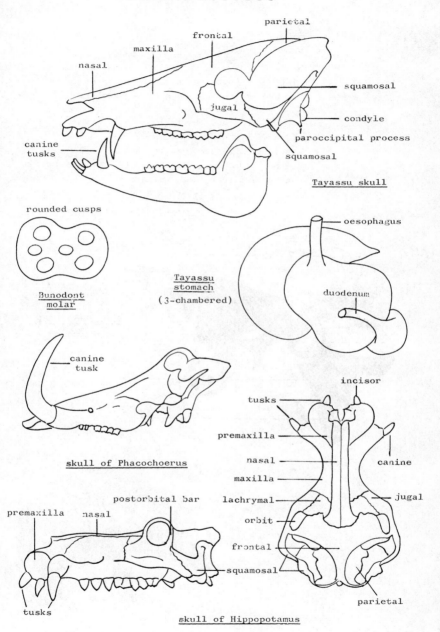

Tayassu skull

rounded cusps

Bunodont molar

Tayassu stomach (3-chambered)

oesophagus

duodenum

canine tusk

skull of Phacochoerus

incisor

tusks

premaxilla

nasal

maxilla

lachrymal

orbit

frontal

squamosal

canine

jugal

parietal

postorbital bar

premaxilla   nasal

tusks

skull of Hippopotamus

ARTIODACTYLES

<u>Family Suidae</u> - Pigs

Suiformes in which:-

1. They are medium-sized to large mammals with
   relatively short legs and coarse, sparse hair.
   They are omnivorous and grub for roots and
   other vegetation and also eat invertebrates
   such as worms and snails.
● 2. The tusks are upwardly turned.   The dental
   formula is variable. (205)
3. They are widely distributed in the Old World
   south of latitude 58°N, but are absent from
   Australasia.   There are 5 genera and 8 species.

Examples:- <u>Sus</u>, <u>Potamochoerus</u>, <u>Babirussa</u>.
<u>Phacochoerus</u>.

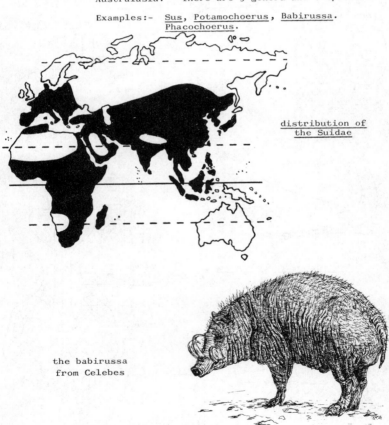

<u>distribution of
the Suidae</u>

the babirussa
from Celebes

A R T I O D A C T Y L E S

Family Tayassuidae - Peccaries

Suiformes in which:-

1. They are medium-sized mammals with relatively short legs, thick coarse hair and a bristly mane. There is a gland at the centre of the back. They are omnivorous and feed on vegetation, small animals and carrion.

2. The tusks point downward and have a sharp cutting edge. The dental formula is I 2/3, C 1/1, Pm 3/3, M 3/3. (205)

3. They are found in tropical America. There is 1 genus and 2 species.

Example:- Tayassu.

distribution of
the Tayassuidae

collared peccary - Tayassu

Family Hippopotamidae - Hippopotamus

Suiformes in which:-

●1. They are large amphibious mammals with relatively short legs and a thick skin with little hair.   The nostrils are closed by valves and raised on protuberances to the level of the eyes and ears thus remaining above water when the rest of the body is submerged.   They feed at night on vegetation at the margin of water to which they return during the day.

●2. Both the incisors and the canines form tusks.(205) The dental formula is I 2-3/1-3, C 1/1, Pm 4/4, M 3/3.

  3. They are found in Africa.   There are 2 genera and 2 species.

Examples:- Hippopotamus, Choeropsis.

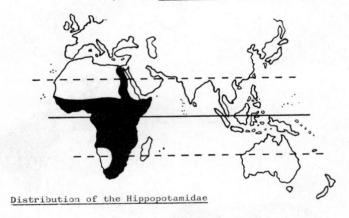

Distribution of the Hippopotamidae

Hippopotamus

ARTIODACTYLES

<u>Suborder Tylopoda</u>, <u>Family Camelidae</u> - Camels and Llamas

Artiodactyla in which:-

1. There are no horns.

● 2. The lower incisors are procumbent and spatulate.
    The canines do not form tusks.   The anterior
    premolars are simple and the remaining cheek teeth
●   have crescent-shaped cusps (selenodont).   The dental
    formula is typically I 1/3, C 1/1, Pm 3/1-2, M 3/3. (210)

3. There is a postorbital bar. (210)

● 4. There is a vacuity between the lachrymal and nasal
     bones. (210)

● 5. The feet have two digits ending in wide pads. (210)

● 6. The metapodials are fused proximally to form a
     cannon bone. (210)

● 7. They ruminate.   The stomach has three chambers. (210)

8. They are medium-sized to large digitigrade mammals (45)
   with relatively long legs and long hair or a fleece.
   The head is held horizontally on a long neck and
   there may or may not be one or two humps on the back.
   They feed on grasses.   Wild forms are found in the
   Gobi Desert and in mountainous regions of South
   America.   There are 2 genera and 4 species.

Examples:-   <u>Camelus</u>, <u>Lama</u>.

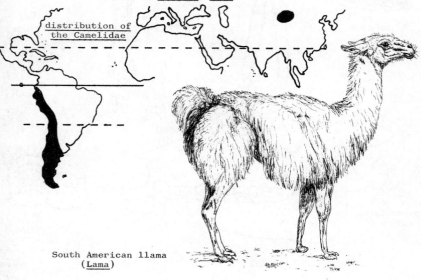

distribution of
the Camelidae

South American llama
   (<u>Lama</u>)

A R T I O D A C T Y L E S

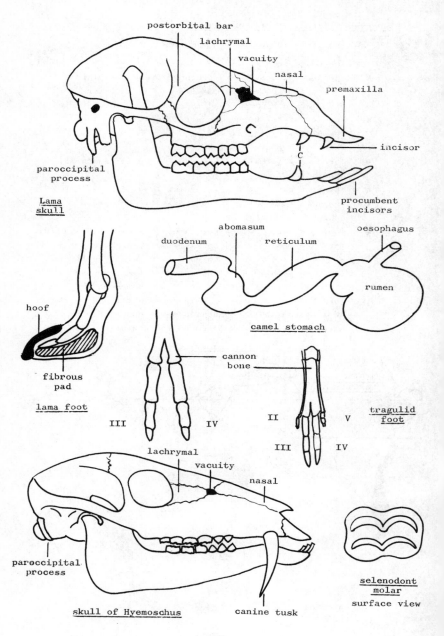

postorbital bar

lachrymal

vacuity

nasal

premaxilla

incisor

paroccipital
process

Lama
skull

procumbent
incisors

abomasum

oesophagus

duodenum

reticulum

rumen

hoof

camel stomach

fibrous
pad

cannon
bone

lama foot

tragulid
foot

III   IV

II   V

III   IV

lachrymal

vacuity

nasal

paroccipital
process

selenodont
molar
surface view

skull of Hyemoschus

canine tusk

A R T I O D A C T Y L E S

<u>Suborder Tragulina</u>, <u>Family Tragulidae</u> - Chevrotains

Artiodactyla in which:-

1. There are no horns.

2. The upper canines form tusks, particularly in the
   male.   The anterior premolars have elongated crowns
   with sharp cutting edges.   The last upper premolar
   ● and the molars have crescent-shaped cusps (selenodont).
   The dental formula is I 0/3, C 1/1, Pm 3/3, M 3/3.

3. There is a postorbital bar.

4. There may or may not be a vacuity between the
   lachrymal and nasal bones.

● 5. The feet have four digits, but the second and fifth
   are short and extremely slender.

● 6. The metapodials are fused proximally to form a
   cannon bone.

● 7. They ruminate.   The stomach has three chambers.

8. They are small, slender-legged mammals with short
   hair feeding on fallen fruit and water plants in
   the tropical forests of West Africa, India and
   Southeast Asia.   There are 2 genera and 4 species.

Examples:- <u>Tragulus</u>, <u>Hyemoschus</u>.

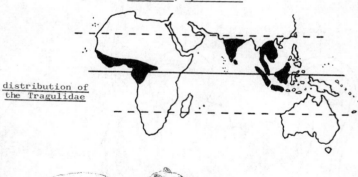

<u>distribution of
the Tragulidae</u>

Malayan chevrotain - <u>Tragulus</u>

211

Suborder Pecora

Artiodactyla in which:-

- 1. Horns are present in the male and in some cases in the female apart from some Cervidae (<u>Moschus</u>) which lack horns.
  2. The upper canines do not form tusks except in those Cervidae without horns.   All the cheek teeth have
- crescent-shaped cusps (selenodont).   The dental formula is typically I 0/3, C 0/1, Pm 3/3, M 3/3. (210)
  3. There is a postorbital bar. (210)
- 4. There is typically a vacuity between the lachrymal and nasal bones. (210)
- 5. The feet have two or four digits, the second and fifth when present being reduced or vestigial. (203)
- 6. The metapodials are fused throughout their length to form a cannon bone. (203)
- 7. They ruminate.   The stomach has four chambers.

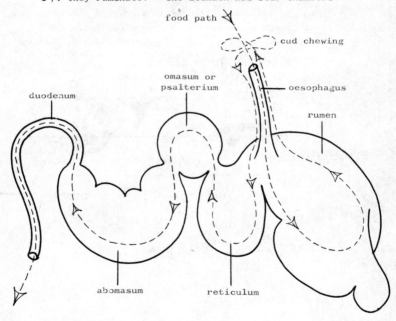

the ruminant stomach

A R T I O D A C T Y L E S

<u>Family Cervidae</u> - Deer

    Pecora in which:-

      1. They are small to large mammals with short hair, feeding on grass, twigs, bark and moss.

● 2. The branching horns (antlers) are normally shed each year and usually present only in the male.

● 3. The feet have four digits.

      4. They are world-wide in distribution except for Africa and Australasia. There are 16 genera and 37 species.

        Examples:- <u>Cervus</u>, <u>Alces</u>, <u>Rangifer</u>, <u>Moschus</u>, <u>Cervulus</u>.

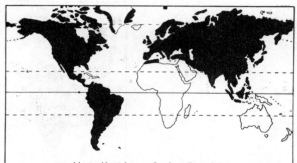

<u>distribution of the Cervidae</u>

reindeer - <u>Rangifer</u>

213

Family Giraffidae - Giraffe and Okapi

Pecora in which:-

1. They are large mammals with short hair and long (okapi) or very long legs (giraffe). They feed on leaves and twigs from bushes and trees.
● 2. Short, unbranched, skin-covered horns, which are not shed, are present in both sexes.
● 3. The feet have two digits.
4. They are found in Africa south of the Sahara. There are 2 genera and 2 species.

    Examples:- Giraffa, Okapia.

distribution of
the Giraffidae

giraffe - Giraffa

A R T I O D A C T Y L E S

<u>Family Antilocapridae</u> - Pronghorn

Pecora in which:-

1. They are medium-sized mammals with short hair, feeding on grass, leaves and twigs.
- 2. The horns are vertical with a forwardly projecting prong halfway along the length and a backwardly hooked tip. They are covered with a sheath of fused hair which is shed each year. They are present in both sexes.
- 3. The feet have two digits.
4. They are found in western North America. There is 1 genus and 1 species.

Example:- <u>Antilocapra</u>.

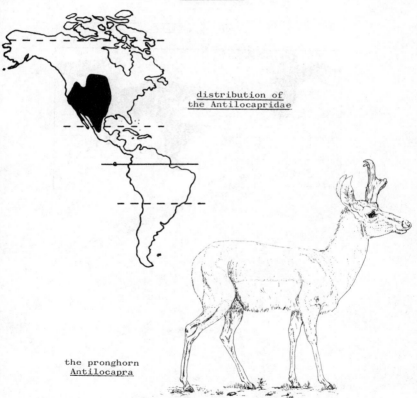

distribution of
the Antilocapridae

the pronghorn
<u>Antilocapra</u>

215

Family Bovidae - Antelopes, Cattle, Sheep and Goats

Pecora in which:-

1. They are small to large mammals with short or long hair and, in the majority of species, with the head directed downward. They feed mainly on grass but other vegetation is eaten.

● 2. Unbranched horns covered with keratin are present in the male and often in the female and not shed.

3. The feet have two or four digits. (203)

● 4. In the wild they are world-wide in distribution except for eastern North America and South America, northwestern Europe and Asia, Madagascar and Australasia. There are 44 genera and 111 species.

Examples:- <u>Bos</u>, <u>Cephalophus</u>, <u>Hippotragus</u>, <u>Antilope</u>, <u>Capra</u>, <u>Ovis</u>.

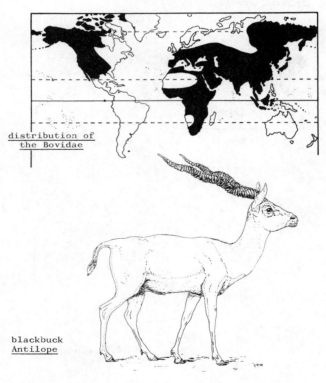

<u>distribution of
the Bovidae</u>

blackbuck
<u>Antilope</u>

12 Glossary

Abomasum - the fourth chamber of the stomach of a ruminant
   artiodactyle. (pages 210, 212).

Acetabulum - the socket in the pelvis that receives the head
   of the femur or thigh bone of the leg. (page 27).

Acromion - a ventral prolongation of the spine of the scapula
   in the mammal, articulating with the clavicle or collar
   bone when one is present. (page 27).

Alisphenoid - a bone forming part of the wall of the cranium
   or braincase. (pages 26, 87).

Alisphenoid canal - a canal running through the alisphenoid
   bone in some Carnivora and Pinnipedia.   It carries a
   branch of the external carotid artery. (page 174).

Allantoic placenta (chorio-allantoic placenta) - a placenta
   formed from the embryonic membranes, the allantois and
   chorion, and the wall of the uterus in eutherian mammals
   and thereby supplying food and oxygen and removing waste
   products from the foetus. (page 41).

Allantois - a sac arising from the posterior region of the
   alimentary canal in tetrapods.   In embryonic reptiles
   and birds it receives waste products.   In eutherian
   mammals it forms part of the placenta. (page 41).

Amnion - the embryonic membrane enclosing a fluid-filled
   cavity containing and protecting the embryos of reptiles,
   birds and mammals. (page 41).

Angular - the bone forming the angle of the lower jaw in most
   gnathostomes.   In the mammals it becomes associated with
   the ear as the tympanic bone. (page 15).

Anogenital - pertaining to the region of the anus and
   genitalia.

Antitragus - the lower posterior part of the pinna of the
   ear opposite the tragus. (page 56)

Aortic arches - the arteries that pass from the ventral aorta
   to supply the gills in fishes and thence unite dorsally
   to form the dorsal aortae.   They become modified and
   reduced in number in the tetrapods. (page 18).

Articular - posterior bone of the lower jaw forming the
   articulation with the upper jaw in gnathostomes except
   mammals where it becomes the malleus of the middle
   ear. (page 15).

GLOSSARY

Astragalus - one of the proximal tarsal bones in the higher vertebrates. (page 27).

Atlas vertebra - the first vertebra of the neck articulating with the occipital region of the skull. (page 16).

Auditory meatus - the canal leading from the exterior to the tympanum or ear drum. (pages 26, 87).

Axis vertebra - the second vertebra of the neck forming a pivot on which the atlas vertebra turns. (page 16).

Baleen - comb-like sheets of keratin hanging down from the roof of the mouth in whalebone whales (Mysticeti) and serving to collect plankton. (page 165).

Brachydont - teeth with short or low crowns and well-developed roots with narrow canals.

Bulla - a hollow, thin-walled, often hemispherical protuberance on the base of the skull below the middle ear in mammals. (page 80).

Bunodont - a tooth having rounded cusps characteristic of the premolars and molars of omnivorous mammals. (page 205).

Calcaneum - the tarsal bone that forms the heel in mammals. (page 27).

Calcar - a process of the calcaneum of a bat helping to support the uropatagium. (page 54)

Canine - a pointed tooth situated behind the incisors and in front of the premolars. (page 26).

Cannon bone - a single pillar-like bone in the distal part of the leg in the more advanced ungulates. It is formed in artiodactyles by the fusion of the third and fourth metapodial bones. In the horses it is the third metapodial. (pages 203, 210).

Carnassials - the last upper premolar and the first lower molar in higher Carnivora cutting with a scissor-like shearing action. (page 174).

Carotid arteries - the chief arteries that pass up the neck and supply the head. (page 18).

Carpals - the bones of the wrist. (page 154).

Catarrhine - having the nostrils close together and directed downward. (page 86).

Cement - a layer similar in structure to bone coating the
part of the tooth embedded in the jaw. It sometimes
forms a thin layer over the enamel on the exposed part
of the tooth. (page 16).

Centrum - the cylindrical body of a vertebra. (pages 17, 95).

Cerebellum - the major lobes of the hind brain concerned
largely with equilibration and coordination of muscle
activity. (pages 16, 42).

Cerebrum (Cerebral hemisphere) - the major lobes of the fore
brain. In mammals it is mainly concerned with the
analysis of information about the environment and the
instigation of appropriate responses. In the highest
forms it is the seat of reason. (pages 16, 42).

Cervical vertebra - one of the vertebrae supporting the
neck. (page 16).

Cheek teeth - the premolar and molar teeth. (page 16).

Chorion - the outer embryonic membrane of reptiles, birds
and mammals. In eutherian mammals it is a component of
the chorio-allantoic placenta. (page 41).

Cloaca - a common chamber into which the alimentary, urinary
and reproductive systems discharge. (pages 20, 25).

Cochlea - part of the inner ear of birds and mammals concerned
with the perception of sound. In mammals other than
monotremes it is spirally coiled. (page 15).

Columella auris - ·a rod of bone connecting the tympanic
membrane and the inner ear in amphibians, reptiles and
birds. It is derived from the hyomandibular bone of
fishes which joins the upper jaw to the cranium. In
mammals it forms the stapes, the inner auditory ossicle
in the middle ear. (page 15).

Condyle - a rounded knob of bone fitting into a corresponding
socket in another bone. (pages 15, 16, 26).

Coracoid - a bone extending from the scapula to the sternum
in reptiles, birds and monotremes, but reduced in higher
mammals to the coracoid process of the scapula. (page 27).

Corpora quadrigemina - two pairs of lobes from the dorsal
surface of the midbrain of mammals, the anterior pair are
optic and the posterior pair auditory centres. (page 42).

Corpora trapezioidea - a band of nerve fibres connecting the
right and left lobes of the cerebellum in the brain. (page 42).

Corpus callosum - a band of nerve fibres connecting the
neopallial region of the right and left cerebral
hemispheres of the brain in eutherian mammals. (page 42).

Cowper's gland - a gland discharging into the male urethra.
(pages 20, 25).

GLOSSARY

Crenulate - with a margin of rounded scallops or indentations.

Crepuscular - active in the twilight (at sunrise and sunset).

Crura cerebri - a band of nerve fibres connecting the cerebral hemispheres and the lobes of the cerebellum on each side of the brain. (page 42).

Cuboid bone - one of the distal tarsal bones in many higher vertebrates. It is approximately cubical and supports the fourth and fifth metatarsals. (page 27).

Cuspidate - teeth bearing rounded prominences or cusps.

Dentary - the anterior bone of the lower jaw bearing teeth. In mammals it forms the entire lower jaw. (pages 15. 16).

Dentine - the bulk of the tooth composed of organic fibrils impregnated with calcium salts. (pages 16, 188).

Diaphragm - a transverse muscular and membranous partition dividing the body cavity in mammals into thoracic and abdominal regions. (page 18).

Diastema - a significant gap between the teeth, notably between the canine and the first premolar. (pages 27, 103, 106).

Didactylous - the condition in the marsupials where the second and third digits of the hind foot are not united in a common sheath.

Digitigrade - where, in walking, the hind part of the foot is raised and only the digits are in contact with the ground. (page 45).

Diphyodont - having two successive sets of teeth, milk or deciduous and permanent.

Diprotodont - the condition in marsupials where the first pair of incisor teeth are enlarged. (page 27).

Dorsal aorta - the main blood vessel supplying the posterior region of the body. (page 18).

Duodenum - the part of the alimentary canal immediately following the stomach. It receives the bile and pancreatic ducts. (pages 205, 210, 212).

Duplicidentate - the condition in the lagomorphs where the first upper incisor is large with a persistent pulp and the second, which is small and peg-like, is directly behind the first. (page 103).

Echolocation - using the difference in time lapse between the
    production of a sound and the receipt by each ear of the
    echo from a distant object to estimate the position of
    that object.

Enamel - hard outer layer of the tooth. (page 16).

Endemic - restricted to a particular area or region such
    as an island or continent.

Epiglottis - a cartilagenous flap serving to close the
    entrance to the trachea during swallowing. (page 16).

Epiphysis - a cap of bone that ossifies separately from the
    main bone. (page 17).

Epipubic bone - a rod-like bone attached to the pubic region
    of the pelvis and directed forward. In monotremes and
    marsupials it serves to support the marsupial pouch. (page 27).

Erythrocyte - a red blood cell.

Eustachian canal - duct connecting the middle ear with the
    pharynx and serving to equalise the air pressure on either
    side of the tympanum. (pages 15, 16).

Exoccipital - the bones on either side of the foramen magnum
    of the skull. In the mammals they form the condyles
    with which the atlas vertebra articulates. (page 26).

Fallopian tube - the duct connecting the ovary to the uterus
    in the mammal. (page 41).

Fenestrated - having one or more openings. (pages 26, 103)

Foramen magnum - the posterior opening in the skull surrounded
    by occipital bones through which the spinal cord passes
    from the brain. (pages 15, 26).

Frontal - a pair of bones forming the roof of the anterior
    part of the cranium or braincase. (page 26).

Glenoid cavity - the articulating socket in the scapula
    receiving the head of the humerus. (page 27).

Glenoid fossa - a smooth shallow depression on the ventral
    surface of the skull receiving the condyle of the lower
    jaw. (page 26).

GLOSSARY

Habenular commissure - a transverse tract of nerve fibres connecting the habenular nuclei or ganglia on each side of the fore brain. (page 42).

Hair - keratin fibres of epidermal origin forming the fur of mammals. (page 14).

Hallux - the first (preaxial) digit of the hind foot; for example the great toe of higher Primates.

Heptamerous - composed of seven parts.

Heterodont - with teeth differentiated into various types such as for biting, tearing, crushing, cutting and grinding. (page 26).

Hippocampal commissure - a transverse tract of nerve fibres connecting the hippocampal regions of each cerebral hemisphere of the brain. The hippocampus forms the major part of the archipallium which evolved before the neopallium and is concerned with olfaction. (page 42).

Humerus - the bone of the upper arm or forelimb. (page 154).

Hypocone - the postero-internal cusp of the mammalian upper molar. The corresponding cusp of the lower molar is the hypoconid and is postero-external. (page 45).

Hypsodont - teeth having high crowns and short roots as in the molars of the horse. (page 198).

Ilium - the dorsal bone of the pelvis articulating with the sacral region of the vertebral column. (page 27).

Incisor - one of the teeth in front of the canine in mammals. In the upper jaw they are borne by the premaxilla. (pages 16, 26).

Incus - the central auditory ossicle of the mammalian middle ear derived from the quadrate bone. (page 15).

Infra-orbital foramen - a canal in the maxilla beneath and anterior to the orbit of the eye carrying nerves and blood vessels onto the face. (pages 106, 192).

Interclavicle - a ventral, median bone in the pectoral girdle lying between the clavicles or collarbones in many reptiles and monotremes.

Interparietal - a small bone present in some mammals between the parietal bones which form the roof of the posterior part of the cranium or braincase. (pages 152, 154).

Ischial callosity - a hard or thickened area of skin on the
    buttock above the ischium prominent in Old World
    monkeys.

Ischium - the posterior bone of the pelvis. (page 27).

Jugal - a bone on the lateral part of the face below the eye
    forming part of the zygomatic arch. (page 26).

Keratin - an insoluble protein rich in sulphur-containing
    amino acids.   It is formed by epidermal tissues and is
    the substance of hair, hoof, horn, scales and feathers.

Lachrymal - a small bone of the face lying within or just
    outside the orbit and usually perforated to carry the
    duct draining tears from the lachrymal gland (when present)
    into the nasal cavity. (pages 26, 80, 87).

Laminate - made of layers of one or more materials.

Larynx - the upper part of the trachea containing the vocal
    cords and modified for the production of sound. (pages 15, 16

Lingual - lying near or next to the tongue.

Loph - a ridge on the grinding surface of a tooth. (page 45).

Lophodont - having teeth with ridges joining the cones to
    form a grinding surface. (page 45).

Malleus - the outer auditory ossicle of the mammalian middle
    ear derived from the articular bone. (page 15).

Mammary gland - milk-producing gland characteristic of
    mammals. (pages 14, 20).

Marsupial pouch - an abdominal pouch formed by a fold of skin
    behind and usually covering the mammary glands in
    monotremes and many marsupials. (page 20).

Masseter muscles - the muscles that raise the lower jaw.
    They pass from the angle of the jaw and are usually
    inserted on the zygomatic arch and the face. (page 106).

Mastoid process - a process of the temporal bone behind the
ear. The temporal bone itself is a compound bone formed
from the squamosal, petrosal and the tympanic bones.
(page 26).

Maxilla - the posterior bone of the upper jaw bearing
teeth. (page 26).

Medulla - the posterior portion of the brain linking the
spinal cord with the higher centres and concerned with
the control of involuntary functions. (page 42).

Mesaxonic - with the weight-bearing axis of the foot passing
along the third digit as in the tapir and the horse.
(page 197).

Metacarpal - one of the five bones of the hand or fore foot
between the carpals and the phalanges. (page 154).

Metacone - the posterior of the three cusps of the primitive
triconodont upper molar. In higher forms where the cusps
are not in line it is the postero-external cusp. The
corresponding cusp of the lower molar is the metaconid
and is postero-internal in position. (page 45).

Metapodium - the collective term for both the metacarpal
and the metatarsal. (page 45).

Metapophysis - an additional anterior articulating facet on
the vertebra of Edentata. (page 95).

Metatarsal - one of the five bones of the hind foot between
the tarsals and the phalanges.

Middle ear - central chamber containing the auditory
ossicles in the mammalian ear. (page 15).

Molar - the posterior teeth in the mammal not preceded by
milk teeth. (page 26).

Molariform row - cheek teeth in which the premolars and
molars are similar and form a continuous row for
grinding. (pages 106, 192).

Monophyodont - having a single set of teeth which are not
replaced.

Nasal bone - one of a pair of bones forming the roof of
the nasal cavity. (page 26).

Neopallium - a new part of the cerebral cortex that develops
in the mammal brain associated with complex learned
behaviour. (pages 16, 42).

Neural spine - the dorsal process of the neural arch of
a vertebra. (pages 17, 95).

Obturator foramen - a large opening between the pubic and
    ischial parts of the mammalian pelvis. (page 27).

Occipital condyle - the rounded condyle(s) in the occipital
    region of the skull that articulates with the atlas
    vertebra. (pages 15, 16).

Occlusal - concerning the biting or grinding surface of
    a tooth.

Odontoid process - a peg-like projection from the anterior
    end of the centrum of the axis vertebra on which the atlas
    vertebra turns. It is formed from the centrum of the
    atlas vertebra. (page 16).

Oesophagus - the alimentary canal between the pharynx and
    stomach. (pages 16, 205, 210, 212).

Omasum - the third chamber of the stomach of a ruminant
    artiodactyle. (page 212).

Orbit - the eye socket. (pages 80, 87).

Otic capsule - part of the skull of vertebrates enclosing
    the inner ear. (page 15).

Palate - the roof of the mouth separating the buccal and
    nasal cavities. (pages 16, 26).

Pallium (Archipallium)- that part of the cerebral cortex of
    vertebrates concerned with olfaction, as distinct from
    the neopallium which evolved later and is found in
    addition in the mammals. (page 42).

Pantropical - throughout the tropics.

Paracone - the anterior of the three cusps of the primitive
    triconodont upper molar. In higher forms where the
    cusps are not in line it is the antero-external cusp.
    The corresponding cusp of the lower molar is the paraconid
    and is antero-internal in position. (page 45).

Paraxonic - with the weight-bearing axis of the foot passing
    between the third and fourth digits as in the pig and
    the cow. (page 203).

Parietal - one of a pair of bones forming the roof of the
    posterior part of the cranium or braincase. (page 26).

Paroccipital process - a downwardly projecting process at the
    back of the skull from or at the side of the exoccipital
    bone and behind the ear. It serves as an attachment
    for the digastric muscle which opens the jaw. (pages 15, 106).

Patagium - in mammals a fold of skin connecting the fore and
    hind limbs which, when stretched, is used for gliding.
    (page 54).

G L O S S A R Y

Pectoral girdle - the skeletal arch that supports the
anterior pair of appendages in vertebrates. (pages 17, 27).

Pelvis (pelvic girdle) - the skeletal arch that supports the
posterior pair of appendages in vertebrates. (page 27).

Penis - the copulatory organ of the male. (pages 20. 25).

Pentadactyl - having limbs with five digits as in the
tetrapods.

Periotic bone - the bone surrounding the ear. Typically it
is composed of three elements, the pro-otic, epiotic and
opisthotic bones, completely fused to form the periotic
in mammals.

Persistent pulp - where the pulp cavity of a tooth remains
open at the base allowing continuous growth of the
tooth. (page 106).

Phalanges - the bones of the digits. (pages 45, 154).

Pinna - a trumpet-like extension of the external ear (often
movable), found only in mammals. It is supported by
cartilage and projects from the general head surface.
(pages 15, 55, 56).

Plantigrade - where, in walking, the entire foot (and hand
in quadrupeds) is in contact with the ground as in Man
and the bear. (page 45).

Platyrrhine - having the nostrils well separated and facing
more or less laterally. (page 86).

Pollex - the first (preaxial) digit of the fore foot, that
is the thumb of higher Primates for example.

Polyphyletic - derived from more than one ancestral line

Polyprotodont - the condition in marsupials where the first
pair of incisor teeth are not enlarged and there are three
pairs or more in the upper and lower jaws. (page 27).

Pons Varoli - a transverse tract of nerve fibres conspicuous
on the ventral surface of the mammalian brain at the
anterior end of the medulla. (page 42).

Postorbital bar - a bar of bone forming the posterior margin
of the orbit and connecting the frontal bone and the
zygomatic arch. It is usually composed of the frontal
and the jugal. (pages 80, 192, 198).

Postzygapophysis - the posterior facet on a vertebra
articulating with the prezygapophysis of the vertebra
behind. (pages 17, 95).

Precoracoid - a bone of the pectoral girdle in many amphibians
and reptiles lying in front of the coracoid.

Premaxilla - the anterior bone of the upper jaw bearing the
upper incisor teeth. (pages 16, 26).

G L O S S A R Y

Premolar - the cheek teeth in the mammal situated between the canine and the molars. They are preceded by milk teeth. (page 26).

Prezygapophysis - the anterior facet on a vertebra articulating with the postzygapophysis of the vertebra in front. (pages 17, 95).

Procumbent incisors - incisors which project more or less horizontally, that is in line with the lower jaw as in the lemurs and the camel. (pages 80, 210).

Propatagium - the region of the patagium anterior to the fore-limb in bats. (page 54)

Protocone - the central cusp of three in the primitive triconodont upper molar. In higher forms where the cusps are not in line it is the antero-internal cusp. The corresponding cusp of the lower molar is the protoconid and is antero-external in position. (page 45).

Psalterium - an alternative name for the third chamber of the stomach of a ruminant artiodactyle. It is also known as the manyplies through the resemblance of the folds of its wall to the pages of a book. (page 212).

Pterygoid - the central element of the palato-pterygo-quadrate bar which forms the upper jaw in elasmobranch fishes. In bony fishes and tetrapods the tooth-bearing premaxilla and maxilla become the functional upper jaw and the palatine and pterygoid are incorporated into the palatal region. In the mammal it is a thin wing-shaped bone. (page 174).

Pubis - the anterior and ventral bone of the pelvis. (page 27).

Pulmonary vein and artery - the blood vessels to and from the heart supplying the lungs. (page 18).

Pulp cavity - the central region of the tooth containing blood vessels and nerves. (pages 16, 188).

Pyramidal tract - a longitudinal tract of nerve fibres connecting the medulla and the cerebellum on each side of the brain. (page 42).

Quadrate - posterior bone of the upper jaw forming the point of articulation with the lower jaw in gnathostomes except mammals where it becomes the incus of the middle ear. (page 15).

Quadrupedal - walking on four feet.

Radius - the preaxial bone of the forearm or corresponding
part of the fore limb of the tetrapods. The postaxial
bone is the ulna. (pages 55, 154).

Rectum - the posterior region of the alimentary canal.
(pages 20, 25).

Reticulum - the second chamber of the stomach of a ruminant
artiodactyle having the mucous membrane folded to form
hexagonal cells, hence its alternative name honeycomb
stomach. (pages 210, 212).

Rhinarium - the hairless area surrounding the nostrils in
mammals. (pages 80, 103).

Rumen - the first chamber of the stomach of a ruminant
artiodactyle. (pages 210, 212).

Rumination - the act of chewing the cud whereby, in ruminant
artiodactyles, food is regurgitated from the rumen into
the mouth and rechewed before passage into the
reticulum. (page 212).

Scapula - the shoulder blade or dorsal part of the pectoral
girdle. (pages 17, 27).

Scrotum - the external sac containing the testes in those
mammals where the testes descend from the abdominal
position.

Sebaceous gland - a fat-producing gland in the skin. (page 14).

Selenodont - having molar teeth with crescentic ridges on
their occlusal surface as in many artiodactyles. (page 210).

Semicircular canals - the organs of balance in craniates, part
of the inner ear. (page 15).

Simplicidentate - the condition in the rodents where there
is a single large upper incisor with a persistent
pulp. (page 106).

Spatulate - with a broad, more or less flattened end like a
spatula or spoon, as in the lower incisor teeth of the
camel.

Squamosal - the posterior bone of the margin of the skull in
higher craniates. It forms the articulation with the
lower jaw in mammals. (pages 15, 26).

Stapes - the inner auditory ossicle of the mammalian ear.
It is derived via the columella auris of reptiles from
the hyomandibular bone of fishes which joins the upper
jaw to the cranium. (page 15).

Sternum (breast bone) - a series of bones or cartilages
lying in the mid-ventral line of the thorax, to which the
ventral ends of most ribs are attached. (page 17).

G L O S S A R Y

Sublingual gland - a pair of salivary glands situated beneath the tongue. (page 16).

Submaxillary gland - a pair of salivary glands situated near the angles of the jaw. (page 16).

Supra-occipital - the dorsal bone lying above and forming the dorsal margin of the foramen magnum in the skull. (page 26).

Supratragus - the upper part of the pinna of the ear. (page 56)

Syndactylous - the condition in the marsupials (and birds) where the second and third digits of the hind foot are enclosed in a common sheath. (page 27).

Systemic arch - the aortic arch leading to the main blood vessel supplying the body from the heart. (page 18).

Tarsus - the ankle, or region between the tibia (and fibula) and the metatarsus.

Temporal fossa - a broad cavity in the mammalian skull behind the orbit mainly occupied by muscles used for raising the jaw. (page 80).

Tetrapoda - 'four-footed' animals comprising the classes Amphibia, Reptilia, Aves and Mammalia.

Thecodont - with teeth inserted in sockets. (page 16).

Trachea - the windpipe through which air passes to and from the lungs. (pages 15, 16, 18).

Tragus - the flap in front of the external aperture of the ear in mammals. It is prominent in some of the bats. (page 56).

Transverse process - the lateral process arising from the centrum of a vertebra. (page 17).

Triconodont tooth - having three simple conical cusps in a line, paracone, protocone, metacone. It is found in the earliest mammals. (page 45).

Turbinal bones - thin scroll-like bones in the nasal cavity covered with mucous membrane. (page 16).

Tympanic bone - the bone that supports the tympanic membrane in mammals. It is derived from the angular bone of the lower jaw. In many mammals it forms the bulla. (pages 15, 26, 87).

Tympanic membrane - the ear drum transmitting vibrations to the auditory ossicles and thence to the cochlea. (page 15).

G L O S S A R Y

Ulna - the postaxial bone of the forearm or corresponding
part of the fore limb of tetrapods. The preaxial bone
is the radius. (pages 55, 154).

Umbilical cord - a cord containing two arteries and a vein
through which blood passes between the foetus and the
placenta. It is formed from the allantois. (page 41).

Unguligrade - walking on the tips of the digits which are
provided with broad hooves. (page 45).

Ureter - the urinary duct draining the kidney into the
bladder. (page 20).

Uropatagium - the region of the patagium extending between
the hind limbs in bats. (page 54)

Uterus - the part of the reproductive tract in the female
mammal in which the embryo is implanted and the foetus
developed. (pages 25, 41).

Vacuity - a gap between the bones of the skull, particularly
in some artiodactyles between the lachrymal and nasal
bones. (page 210).

Vagina - the canal(s) which leads from the uterus in the
female mammal to the cloaca or to the exterior. (pages 25,41)

Vas deferens - the sperm duct from the testis. (page 20).

Vena cava - the principal vein(s) draining blood from the
posterior region of the body to the heart. (page 18).

Yolk sac - a sac arising from the alimentary canal in the
embryos of reptiles, birds and mammals. In the reptiles,
birds and monotremes it contains the yolk which nourishes
the embryo, but in all other mammals it does not. In
the marsupials the yolk sac absorbs nutrient fluids
produced by the uterine wall and in one case forms a
short-lived placenta. (page 41).

Zygoma - see zygomatic arch.

Zygomatic arch - a bar of bone extending along the side of
the skull beneath the orbit in most animals. It is
composed of the maxilla, jugal and squamosal, but may
be incomplete. (page 26).

List of generic names quoted

Chrysochloris (Cape golden mole) 48*

Civettictis (African civet) 179

Coendou (tree porcupine) 133*

Colobus (colobus monkey) 90

Craseonycteris (Thailand bat) 59

Cricetus (hamster) 120

Crocidura (shrew) 50

Crocuta (spotted hyaena) 180

Cryptomys (mole rat) 146

Ctenodactylus (gundi) 147*

Ctenomys (tuco-tuco) 141*

Cuniculus (paca) 137

Cyclopes (dwarf anteater) 96

Cynocephalus (flying lemur) 53*

Cystophora (hooded seal) 186

Dactylomys (arboreal rat) 143

Dasyprocta (agouti) 137*

Dasypus (common armadillo) 95* 98

Dasyurops (marsupial cat) 29, 30*

Dasyurus (marsupial cat) 12, 23, 29

Daubentonia (aye-aye) 80, 82, 83*

Delphinapterus (white whale) 160

Delphinus (dolphin) 161

Dendrohyrax (tree hyrax) 191* 192*

Dendrolagus (tree kangaroo) 37, 38*

Dendromus (African climbing mouse) 124

Desmana (desman) 51

Desmodus (vampire bat) 70*

Diaemus (white-winged vampire bat) 70

Diarthrognathus 15

Diceros (African rhinoceros) 199

Diclidurus (ghost bat) 60

Didelphis (opossum) 5, 28*

Dinomys (pacarana) 136*

Diphylla (hairy-legged vampire bat) 70*

Dipodomys (kangaroo rat) 114

Dipus (Egyptian jerboa) 129*

Dolichotis (mara) 134

Dugong (dugong) 192* 194*

Echimys (crested spiny rat) 143

Echinosorex (moon rat) 49

Eidolon (straw-coloured fruit bat) 55

Elephantulus (elephant shrew) 52

Elephas (Indian elephant) 189

Emballonura (sheath-tailed bat) 60

Enhydra (sea otter) 173, 177

Equus (horse, ass, zebra) 201*

Erethizon (New World porcupine) 133

Erinaceus (common hedgehog) 49*

Eschrichtius (grey whale) 165, 166*

Euchoreutes (long-eared jerboa) 129

Eumops (mastiff bat) 77*

Felis (cats) 181*

Fossa (Malagasy civet) 179

Funambulus (palm squirrel) 111

Furipterus (smokey bat) 72

Galago (bushbaby) 84

Genetta (genet) 179

Geocapromys (short-tailed hutia) 139

Geomys (pocket gopher) 113*

Georychus (blesmol) 146*

Gerbillus (gerbil) 120

Giraffa (giraffe) 214*

Glirulus (Japanese dormouse) 125

List of common names quoted

Indian civet - <u>Viverra</u>